The Open World and Closed Societies

The Open World and Closed Societies
Essays on Higher Education Policies "in Transition"

Voldemar Tomusk

THE OPEN WORLD AND CLOSED SOCIETIES
© Voldemar Tomusk 2004
Softcover reprint of the hardcover 1st edition 2004 978-1-4039-6507-3
All rights reserved. No part of this book may be used or reproduced in any manner whatsoever without written permission except in the case of brief quotations embodied in critical articles or reviews.

First published 2004 by
PALGRAVE MACMILLAN™
175 Fifth Avenue, New York, N.Y. 10010 and
Houndmills, Basingstoke, Hampshire, England RG21 6XS
Companies and representatives throughout the world

PALGRAVE MACMILLAN is the global academic imprint of the Palgrave Macmillan division of St. Martin's Press, LLC and of Palgrave Macmillan Ltd. Macmillan® is a registered trademark in the United States, United Kingdom and other countries. Palgrave is a registered trademark in the European Union and other countries.

ISBN 978-1-349-52868-4   ISBN 978-1-4039-7947-6 (eBook)
DOI 10.1057/9781403979476

Library of Congress Cataloging-in-Publication Data
Tomusk, Voldemar, 1962–
    The open world and closed societies : essays on higher education policies 'in transition' / Voldemar Tomusk.
        p. cm. (Issues in higher education)
    Includes bibliographical references and index.
    ISBN 978-1-4039-6507-2
    1. Higher education and state—Europe, Eastern—Cross-cultural studies.
    2. Educational change—Europe, Eastern—Cross-cultural studies. I. Title.
    II. Issues in higher education (New York, N. Y.)

LC178.E852T66 2004
378.47—dc22   2003053654

A catalogue record for this book is available from the British Library.

Design by Newgen Imaging Systems (P) Ltd., Chennai, India.

First edition: April 2004
10 9 8 7 6 5 4 3 2 1

Transferred to Digital Printing 2009

To the Memory of My Dad

We are all swallowed up in politics; there is no room for letters; and it is feared that the next generation will not only inherit but improve the polite ignorance of the present.

Simon Ockley, clerk in Holy orders and Sir Thomas Adams's
Professor of Arabic in the University of Cambridge
1678–1720

Contents

Acknowledgments   ix

Preface   xiii

Chapter 1. From Lenin to Digital Rapture:
The Everlasting Transition in East European
Higher Education and Beyond   1

Chapter 2. Higher Education Reform in Romania:
Knocking on Heaven's Door   19

Chapter 3. Thirteen Years of Higher Education Reforms
in Estonia: Perfect Chaos   35

Chapter 4. Russian Higher Education After Communism:
The Candy Man's Gone   53

Chapter 5. Market as Metaphor in East European Higher Education   75

Chapter 6. Reaching Beyond Geometry—
The Privateness of Private Universities   89

Chapter 7. Exploring the Limits of Entrepreneurial Response   105

Chapter 8. When East Meets West: Decontextualizing the
Quality of East European Higher Education   119

Chapter 9. Reproduction of the "State Nobility" in Eastern Europe:
Past Patterns and New Practices   131

Chapter 10. The Communication Community and the Scam of
the Knowledge Society   145

Chapter 11. Transnational Capitalist Class and World Bank
"Aid" for Higher Education   167

Chapter 12. Toward a Model of Higher Education Reform in
       Central and East Europe   183

Epilogue: The Unholy Trinity of Prince, Prophet, and Philosopher   199

Notes   203

Bibliography   223

Index   235

Acknowledgments

It is more than a decade since I wrote my first paper on higher education "in transition" and presented it to an audience of around two people at the annual conference of the Society for Research into Higher Education at the University of Surrey in England. Now, it seems a less than naïve attempt to grasp a process that had not yet even begun. I have been lucky enough being able to spend the intervening decade observing the reforms in East European higher education or, as has often been the case, the nonreforms, as a young faculty member, a government official, and for the past seven years as an open society activist. It has allowed me to experience changes within the university and the Ministry of Education, as well as to see how these appear from the perspective of a major international donor organization. The opportunity to be at the heart of change has been a major learning experience for me, as it has been the opportunity to work with a number of special people who have shaped our world as it enters the new millennium.

I have to confess that there is one single reason why I accepted the position I was offered back in 1995 at the Ministry of Culture and Education of the Republic of Estonia when I returned from a year-long training program at the International Institute of Educational Planning in Paris. It may sound utterly selfish, and it perhaps is, but the only reason for my entering the Ministry was that I wanted to know what was going on. It was a difficult job, and the salary accorded a young family with two small children a living standard just below the poverty level, but I wanted to know what was going on because for somebody looking in from the outside, the steps the Ministry was taking appeared somewhat less than well thought through policy. It did not take long to understand the internal mechanisms involved in that. Neither did much time pass before I was offered an opportunity to apply my newly gained knowledge on a wider international scale in the Open Society Institute in Budapest. It seemed to be another good opportunity to learn and perhaps to do some worthwhile things along the way. So in less than two months from the point at which I had been told about the job, my family and I had settled in Budapest.

The results of my dozen years of practical studies of the East European higher education transition are presented in this book. One might argue, and rightly so, that not a great deal has been accomplished here. It is by no means a systematic study, its theoretical background does not reach the necessary depths, and the empirical evidence is less than complete. If there is anything with which I might excuse myself then perhaps it is that there is no part of this book that was written for professional research purposes. The essays collected here were written at nights and weekends in the belief that recording some of these things was better than not doing so, and that this would enhance our understanding of some of the most significant events of the twentieth century: the fall of the state-socialist political systems in Eastern Europe and the following reforms toward achieving normality, whatever that might mean. During the past few years, I have met with several people along the way whose views really matter to me. Having read a portion of the work collected in this book they encouraged me to come up with a more substantial study. I would like to thank Guy Neave, S. Frederick Starr, and William Weary for their encouraging words without which completing this work would have been considerably more difficult. But by no means should they be held accountable for any deficiencies in the outcome, which is my responsibility only. I would also like to thank Michael Webb for his assistance bridging the gap between the grammar of the English language and my apparently utter disrespect towards it.

One issue I have become aware of as I worked on this book is the responsibility the author has in not abusing those who buy a book and spend moments of their finite lives reading it. I offer my apologies to those who have taken up this book and not found what they had hoped for, still, I hope that there are some people out there whose knowledge and understanding benefits from the days I have spent working on it. My apologies also go to my family, to my wife Anu and our four children Erik, Karl, Miriam, and David who only see me either leaving for the airport, or hiding in my study. They certainly deserve more than that.

When I started my journey in the study of the higher education world, it was a very different place. I still carried the passport of a once-mighty country—the Soviet Union—in my pocket, and traveling with it caused me a great deal of trouble indeed. In a way these were also fascinating days when many things happened and before our very eyes—as a political scientist would say—a new world order was being founded. When the old rules were no longer valid and the new ones were not yet established, events occurred, stories of which would be told for years. In one such story I have heard my colleagues telling I recognize myself as the main actor. It is about how Voldemar, citizen of the Soviet Union from Estonia once entered the United Kingdom without a British visa. Those who have ever experienced passport control at Heathrow Airport in London know perfectly well that such things never happen. But the fact is that sometimes they do. This is a rather unusual story, and to put the records straight I will share it here in writing. For the first and the last time, just to give the reader a little idea of what life was about in the early days of the transition to open democracies. With this, I would also like to thank an anonymous immigration officer who, perhaps unwittingly made a major contribution to my knowledge of higher education.

On January 1, 1992 I was going through a bunch of papers I had collected at a conference I had been to just before that Christmas. Among them I found a promotional leaflet about a course launched by the CHER (The Consortium of Higher Education Researchers), The European Higher Education Advanced Training Course. It was a modular course, consisting of eight weeklong modules in eight European countries over a two-year period. It looked like an attractive opportunity, and so I filled in the form and mailed it to the required office, of which I knew nothing at that moment but learned much later, the Center for Higher Education Policy Studies at the University of Twente.

I mailed the papers and almost forgot the whole affair, which looked too good to ever become true anyway. Until a month later, somebody called me to the university from Finland. This person, who identified himself as Kari Hyppönen of the University of Turku, told me that although I was, because of my country of origin, not eligible to attend the course that was supported by the TEMPUS program, the Finnish government was offering me financial support to participate—to cover my travel, the cost of the course, accommodation, per diem—everything.

I had about four weeks left to wait before the first module was to start at the Anglia Polytechnic in the United Kingdom, and I sent my visa application to the British Embassy in Helsinki, which was closest to Tallinn at that time. When the day came, I left for London via Helsinki to pick up my visa on the way, only to learn in Helsinki that four weeks had not been enough for the British Foreign Office in London to authorize the embassy in Helsinki to issue me a visa. It was Friday afternoon and the embassy was already being closed as I stood there waiting for the visa, which never arrived. A clerk behind the counter, perhaps reading my thoughts, warned me strongly not even to try to fly to London without a visa, "You will be returned immediately."

These were the days, however, when the country whose passport I still held no longer existed. In addition to the Soviet passport I also carried a document commonly called as the "Rumessen-passport." It was a nonrecognized ID card issued by an Estonian separatist movement to people who claimed to be citizens of the Republic of Estonia, named after a dissident intellectual who had launched the initiative of recording the descendants of the Estonian citizenry as it had been on June 17, 1940, the day when Soviet troops invaded Estonia. While the clerk behind the counter had warned me not to fly, another officer passing advised that in the unlikely event that I still decided to fly, I had better show this other, nonvalid Estonian ID card on my arrival. The reason for this was that three countries, the United States of America, the United Kingdom, and Denmark had never accepted the incorporation of Estonia into the Soviet Union, and as a citizen of Estonia I had the right to enter the United Kingdom without a visa under a pre–Second World War agreement.

So, I went to Helsinki airport and, for some strange reason, was allowed to board the plane without a visa. I still remember my thoughts whilst flying to London. First, I wondered if my arrival at Heathrow would turn into a political scandal. That looked rather unlikely for somebody as unimportant as myself, a mere university lecturer. Then, while enjoying my meal, I thought that the worst thing that might happen would be eating the same meal again a few hours later. This gave me some comfort.

On arrival at Heathrow, the first thing to do was to look for the most friendly looking immigration officer. One of the ladies looked rather nice and I approached her with my nonrecognized ID card. She took it, studied from both sides, and asked if I had any other documents with me. "Yes," I said, "I have this Soviet passport, but this no longer reflects the country I am living in. I am an Estonian citizen." And I produced the red Soviet passport. She took and studied that too and asked, to my great good fortune, the right question. She did not ask why I did not have the British visa in my Soviet passport, instead she asked why did not I have an Estonian passport. That was an easy one. "Look, throughout the entire post World War II period the only office that has been issuing Estonian passports is the Estonian consulate in

New York. How do you think I would have been able to go there to get one?" So, she took all my invitation letters and papers and a form she had at hand and started filling it in. This was a 30-day visa. "What is your height?" she asked. "1.85," I responded. "And the color of your eyes?" "Blue," and I took off my glasses to show her that this was actually true. Having written it all down she gave me a piece of paper, a permit to stay in the United Kingdom for 30 days. On the other side Prof. McNay was already getting ready to face the British authorities in an emerging refugee crisis. A few hours later I met with many people who have now become my friends and colleagues, and I told the story for the first time.

This was the beginning of my journey in higher education and higher education policy studies. Many people have supported me on the way, teaching, encouraging, and criticizing. It is impossible to list all the friends and colleagues to whom I feel greatly indebted for being able to complete this work. Kari Hyppönen from Turku, without whom I would not have been able even to begin. Osmo Kivinen is a sociologist from whom I have learned a lot. There is another educational sociologist, Len Barton, whose conferences every year in January in Sheffield have made a major contribution to my understanding of education within a complex social world. Guy Neave from the International Association of Universities and CHEPS has helped me to understand the meaning of higher education policy and its difference from traditional social research, and his encouraging words mean more to me than he imagines. My old friend Glenn Paige from the Center for Global Non-Violence on Hawaii, formerly a professor of political science at Princeton, has helped me to understand wider horizons of intellectuality. Finally, there is a very special woman I would like to thank, Mrs. Deirdre Brennan from SRHE, who welcomed me in her home in London in the days the world was turning upside down and people were being killed on the streets in the country in which I lived. Many years later, I can still recall her knocking on my door early in the morning and coming in with a huge cup of tea.

There are organizations and institutions without whom writing this book would have been impossible. The Open Society Institute in Budapest has been providing me with my daily bread and challenging employment, both of which have been of critical importance in this work. The University of Turku has supported me in various ways for long time, but most recently by its support in completing my doctoral studies in educational sociology. I would also like to thank the Swedish International Development Authority for its generous support for my studies at the IIEP in Paris in 1993/94.

Writing this book has been an exciting process. I would like to share it with the friends and colleagues that I have met during my journey. I believe that they will agree that the pleasure of friends walking and talking together cannot be reduced to anything else. This is part of our common nature as human beings, and in these days of great insecurity I hope that we do not surrender it too easily.

Voldemar Tomusk
Budapest
October 1, 2002

Preface

During a relatively short period between 1989 and 1991 the state-socialist regimes in Europe fell apart. The political and economic model that had of late claimed to represent the future of all mankind all of a sudden ceased to exist. It appeared that the entire "Second World" had been no more than a bluff and Potemkin's village.

As soon as the Soviet Union had ceased to threaten its Warsaw Pact allies with military intervention for not following the Moscow line, all of them decided to go it alone. Even if retribution was quick to follow and the erstwhile friends were asked to pay the full world market price for Soviet oil supplies, it did not bring them back into line. Regimes that had lost their only source of legitimacy—Soviet military power—fell and were replaced with more or less democratically elected governments. Finally, in August 1991 during an unsuccessful coup when Soviet hard-liners tried to reestablish the totalitarian regime, which according to their view had been dangerously weakened during the years under Mikhail Gorbachev, the then secretary general of the Soviet Communist Party, and his *perestroika* policy. The country did not have an effective government for two days. This was sufficient for its 15 Republics to walk each in a different direction. What had happened up to this point was perhaps more than anybody would have expected just a few years earlier. It exceeded the wildest dreams of even the most radical dissidents—complete freedom from the Soviet Union and the great evil itself having fallen apart. For many this would have been the end of history after which 10,000 years of happiness could follow. In reality it was only the beginning of a process now referred to as the "transition."

Instead of leaping forward, East European state-socialist countries entered a long and painful transition toward a goal that is usually described as democratic, liberal, free market capitalism. Over the past decade a large number of volumes have been written in an attempt to analyze the process of devolution—at least from a Marxist point of view—from communism to capitalism. It has transpired that introducing market mechanisms in deeply corrupt societies organized around the social bounds established within the communist party and its youth league is by no means an easy task. Several such studies indicate that former elites have been successful in exchanging their social capital for the material values of the capitalist order. In a way, with the exception of a small number of top party leaders, those who were privileged before are still so, and the burden of the transition is carried by those who had not benefited much before either. This latter point has had the most significant implications for post state-socialist East European higher education and its developments in

1990. Unfortunately not enough attention has been paid to the processes of social reproduction in post state-socialist higher education.

Establishing market economies has been the highest priority goal in East European reforms throughout the 1990s. Results region-wide are mixed. There are countries that have tried and to a significant extent succeeded. There are those that once tried and then gave up and also those that still try to combine the "positive elements of both systems"—a challenging task indeed. But there are also those who are still thinking. Life, however, is not only about economies. Struggling with the harsh realities of everyday life in the transition economies, we often tend to forget that some dozen years ago freedom was more important for many of us than the blue jeans or Scotch whisky that the *nomenklatura* was able to purchase from their hard-currency shops or bring home from their business trips to enemy countries. Freedom was a sweet word, and it still is.

Surprisingly enough, we have learned that it is not only the despotic regimes that constrain our freedom. When in the old days, East European scholars were not able to exercise their freedom to travel, publish, or hold discussions with colleagues abroad because of political control, abolishing the political control has not always brought much freedom. Severe economic difficulties have set a limit on the extent to which freedom can be exercised in many countries. Universities that lack the funding to pay for their telephones and electricity cannot spend on conferences and expensive foreign literature. In a way, it is increasingly difficult to be both poor and smart in the global capitalist society. Becoming smart is very expensive, which sets another limitation upon our freedom, an internal one. To maintain their former status, albeit illusory, universities and individual faculty members choose to ignore the outside world and justify doing so by using political and cultural reasons. It takes a certain amount of courage to step out of the prison. Developments in East European higher education in the 1990s indicate that many choose to stay in the cell even if the door is open. This is particularly unfortunate when there are those who use their administrative or symbolic authority to prevent others from leaving.

An open society cannot be established by simply opening a closed one. Becoming a part of the open world has its economic prerequisites, particularly within the global capitalist order. It also has its psychological dimension, people should be prepared to live in openness. Many in Eastern Europe are talking about open societies nowadays although few have seen one. The most ironic effect of some of the programs launched to promote open societies is that they seem to be creating their own closed networks, exclusive non-governmental organizations that are more profit-seeking than many of the local business organizations, clientele of international aid programs, and foreign foundations, and so on and so forth. Perhaps, the goal of establishing a global open society belongs to the group of utopias described in chapter 1 of this book. Moreover, if such a goal is achieved one might really wonder if it is in fact an open society, or something else. There is no longer an alternative to compare it with. Establishing a global open society as an ultimate goal may too easily turn into its own opposite. For the best of our future and our children we should perhaps maintain the right to have our closed societies as long as we want them, and define our identities within those. There is always a risk that we make a wrong choice as a great nation did in 1917. Then there is big trouble for everybody. But any ultimate solution, even the one that looks like ultimate good for everybody, is infinitely more dangerous. Describing the final world of universal happiness is a religious exercise. As long as we accept ourselves and other people as humans and respect their fundamental right to live, we can always find ways to overcome our differences without washing them away. The horrible events in September 2001 indicate that mankind is quite far even from the modest goal of accepting that all humans have the right to live.

This book brings together a selection of papers on East European higher education reforms. It has been my interest to look at higher education in a broader social context, and I think that there is a strong reason to do so. Higher education reform as it is taking place in Eastern Europe, is largely guided by external agencies that are usually highly technocratic in their approaches. We have consultants flying in, telling us with their charts and tables that enrolments should be raised, staff laid-off or institutions merged. With such manipulations they believe that Eastern Europe can reach OECD standards. What they usually miss is that behind each ratio or figure there is a social world, lives people are living, and that putting the number first most probably will not make any of the societies in Eastern Europe function like those of France, Germany, or the United Kingdom. I am convinced that to save East European higher education, and I regret to say it, but that is exactly what it needs—to be saved, policy analysts should understand complex connections between society and its learning institutions. This is particularly so when society is changing and higher education has lost its relevance. It is my modest goal with this book to open such a direction in East European higher education research.

In chapter 1 of this book I look at some of the radical ideas that have shaped East European higher education in the past, as well as those that are expected to do so in future. It concerns two utopias in particular—the Marxist utopia and the ultimate solution it offered to mankind, and the latest techno-romantic utopia that makes similar or even more far-reaching promises to save humans from the inevitable misery of their physical existence through advances in science and technology. I argue that the latest utopia offers an attractive illusion of the possibility of leaping into the information society without restoring the traditional sectors of the economy and without restoring the intellectual function of the university. Such thinking itself, it is argued, is a sign of extreme weakness of the social science faculties in Eastern Europe. Neither does the "knowledge" from the cable replace the need for local intellectual capacity.

The next part of the book, chapters 2–4 presents three national cases. In these chapters various issues of higher education reforms in Romania, Estonia, and Russia are discussed. In each country I compare legal regulations and a selection of official reports with other evidence available on the status of higher education. In all cases, there appears to be a significant gap between the "official knowledge" as the systems are producing it and the real situation of higher education. In each case, an attempt is made to explain the reasons behind such discrepancies, looking at certain historical, cultural, and political determinants.

Chapter 5 looks into the various forms of markets that have been moving into East European higher education. Here, development has been rather chaotic. In many countries prerequisites, particularly legal regulations, for markets to move in to higher education have been inadequate. This particularly raises concerns about the gap between students' expectations and the level of education that universities are able and willing to provide. While it is often argued that new private universities do not deliver services at the level they advertise, there is no good reason to believe that public universities that are admitting growing numbers of fee-paying students reach the promised standards either. While education provided in public, as well as private higher education institutions in Eastern Europe, may both remain inadequate, it is not the market that chooses the fittest for survival, but the visible hand of the state bureaucracy.

Chapter 6 discusses the reform of higher education from the perspective of Burton Clark's classical triangle model and reinterprets it according to recent developments in Eastern Europe. After that, I look more precisely at the meaning of the "private" higher education institution and conclude, following a division proposed by Levy, that higher education privately funded and governed, does not necessarily carry a private mission. To illustrate this argument, an example of an East European private university is offered.

In chapter 7 I discus another fashionable issue in higher education, entrepreneurialism and its implications in Eastern Europe, following the recent well-known study by Burton R. Clark. Using the example of a private graduate school in one of the East European countries, I argue that orienting toward a particular kind of market—that of the foreign donors—an entrepreneurial university in Eastern Europe can completely ignore the local realities as well as basic efficiency indicators. It can develop nothing short of complete academic irrelevance as long as donor agencies are happy, but certainly not beyond that.

Chapter 8 is devoted to the issue of quality. It looks at the ways in which some *Western* practices, particularly those related to quality assurance, are transplanted in Eastern Europe and through this process, significantly modified. It is argued that East European higher education is not in a position to establish adequate quality standards for itself. First, because it would not be in a position to reach them; and second, that as many of the post state-socialist countries are having economic difficulties, these societies are not in a position to articulate clear expectations for higher education. The result is largely about defining goals for higher education in terms divorced from the realities of local societies, to the extent that they can be considered in terms of political rhetoric, which has little use as guidance for running institutions of higher education.

Chapter 9 discusses issues related to the reproduction of elites in East European higher education. It is argued that the traditional sector of elite training, the engineering school, is losing its importance. The pragmatic party technocrats, who moved into power positions with the revolutions of the late 1980s and secured their positions with the successful privatization of state assets, are now looking for different kinds of educational paths for their offspring. Law and business training may have an important role to play in this. However, there seems to be strong competition rising among the institutions trying to assume control over the field of elite reproduction. While public institutions move to this field with the support of respective state agencies, private schools legitimize themselves through various international (accreditation) schemes and foreign sources.

One of the primary issues in East European higher education, as well as higher education more widely is the social relevance of higher education. As I argue in chapter 10, higher education is caught in the middle of a conflict where academics receive their status internationally, while the research and education they provide should be locally relevant. A mutually acceptable balance is increasingly difficult to maintain under a new global regime where knowledge is produced in a limited number of global centers and only distributed through peripheral universities. This has generated opposition, for example among African intellectuals, many of whom argue for a radical return to traditional local epistemic strategies. In this book, I argue for a more moderate approach, to find a balance where scholars could be relevant both globally and locally. Chapter 11 follows the same issue of globalization, demonstrating how aid provided to support higher education reforms by global institutions with the help of global professionals may instead of development lead to embezzlement of the "recipient" countries' own resources.

Chapter 12 seeks a general model arguing, following the work by Elster, Offe, and others, that East European higher education has had three main reform targets, distant past models, the Humboldtian model, that of Western countries in particular, and the recent past. After an unsuccessful attempt to restore the Humboldtian model on a massive scale and the realization that current Westernization projects do not meet East European expectations, the systems are increasingly returning to their recent past, reinterpreting it in romantic terms. Thus, higher education systems largely lack positive policy agendas and are drifting against the current of change. While a part of that is to do with the problem of East European universities being unable to change themselves from within, it is also about the West that has not been able to offer the necessary intellectual support to higher education in "transition."

CHAPTER 1

FROM LENIN TO DIGITAL RAPTURE: THE EVERLASTING TRANSITION IN EAST EUROPEAN HIGHER EDUCATION AND BEYOND[1]

Behold, I see the heavens opened,
and the Son of Man standing at the right hand of God.

Acts 7:56

Over the past 200 years, *Homo sapiens*, a species of mammal, has made unprecedented progress in improving the quality of its life on the planet Earth. There is no ground whatsoever for the argument that any other member of the animal kingdom— even *Canis familiaris* or *Pongo pygmaeus*—has ever achieved anything even close to the achievements of Homo sapiens, wise man. Its existence is no longer limited by the abilities of its body as a whole or by its organ of information processing, the brain. It has learned how to artificially extend its physical as well as mental faculties. Within a matter of hours it can now move from one continent to another; it can even leave the planet for the space stations it has placed in orbit around the planet. Space probes have been sent, not only to other parts of the solar system, but even beyond it. Having extended its physical capabilities, Homo sapiens has expanded its mental faculties and learned how to store and make available to the entire population within seconds all the information the wise men have ever created, as well as how to communicate with each other in real time. Although, for the time being, not everyone has been connected to everyone else, the most intelligent representatives of the species have already figured out smart ways to get everybody wired within a few years, including by wireless connection. Men have started preparing to replace their naturally evolved carbon-based organic bodies for more durable and compact silicon-based bodies that would allow them to leave the planet and discover the outer reaches of space, searching for other microchips like themselves. It is anticipated that for such an achievement, several of the smartest men will be awarded the Nobel Prize.

The past 200 years are perhaps the most remarkable period of the 100,000 years of history of this one remaining human species. While all the other strains like *Homo habilis* and *Homo erectus* became extinct long ago, the youngest representative of the genus *Homo* has taken significant control over both its environment and bodies. It is not easy to explain such a development. Some may well blame it on evil spirits who have tricked mankind into self-destruction; others might argue with equal success that God's grace has finally fallen on the final work of His hands. Whichever is the case, the expansion of the knowledge man has been able to create and put to use during the past 200 years is more than remarkable. Creating a better world is an idea that guides many of its activities. However, it is not always clear what *better* means in the context of human life.

Unlike other representatives of the animal kingdom, wisdom has allowed man to understand the misery of his existence. Whether good or bad, the life a human lives always ends in the same way—sickness and death. Prince Siddhartha Gautama, a.k.a. Buddha, has summed up human destiny within a simple statement—*Everything is suffering*. For Homo sapiens, the wise man aware of his wisdom, this is but crude injustice.

Marx, Lenin, and Their Utopia

Great minds have always been dreaming of better worlds. Still, not much good has come of it. Ultimately, their dreams have all remained "nowheres," utopias never destined to materialize. At times, visionary leaders have been able to mobilize massive support behind their programs—breaking the continuity of history and together with it the yoke of the curse of original sin that made man mortal and obliged him to sweat for his daily bread, and woman to undergo childbirth in great pain. Yet, despite all the efforts man makes to discontinue history, everything quickly returns to its old track or even worse. If there is anything one can learn from the recent experience of establishing brave new worlds, it is that they always stem from the old and not so brave ones and return to them. Suffice it to remember how Trotsky tried to discontinue government, replacing ministers with a brand new breed, the People's Commissars: "The title of 'Ministers' and 'Council of Ministers' was highly unpleasant to Lenin because it recalled the bourgeois Ministers and their role. Trotsky suggested the term 'people's commissars.' Lenin was thrilled by this. He found the expression 'terribly revolutionary' and the name was adopted. A new language was being created and, so it was thought, new institutions along with it.... But on the intermediary level, which was to prove decisive, that of the institution in its second-order symbolic summit of an administrative apparatus distinct from those who are administrated—on this level, in fact ministers remained, cast in the mould already created by the kings of Western Europe since the end of Middle Ages."[2]

At the peak of the Enlightenment, Karl Marx discovered what he and multitudes of his followers considered to be the ultimate law of the evolution of matter. Not completely unlike some contemporary anthropic principles' cosmologists who put the cart before the horse by arguing that "The Universe must have those properties which allow life to develop within it at some stage in its history"[3] (The Strong Anthropic Principle) or even further, as in: "Intelligent information processing must come into existence in the universe, and, once it comes into existence, it will never die out"[4] (The Final Anthropic Principle). The Marxists also have such a high opinion of themselves that they conclude the necessity of the rest of the universe from their own existence and the brilliance of their revolutionary theories. For Marx in particular, one can easily conclude that, according to his theory, his own birth was a necessity written, if not in the stars then at least in the objective laws of the evolution of matter. But not only that. As his role was to discover the ultimate law of nature in his own birth, it was the culmination of the entire preceding history of the universe. Applying a trick called dialectics, Marx discovered the inevitability of a society that in a most radical way breaks the earlier continuities—simply by *overcoming itself*. He discovered the inevitability of the communist society "in which all resistance, all opaqueness would be absent; a society that would be purely transparent to itself; in which everyone's desires would spontaneously harmonise with everybody's else's, or, in order to harmonise would require merely an airborne dialogue which would never be weighted down by the gum of symbolism; a society that would discover, formulate and realise its collective will without having to pass through institutions, or in which institutions would never pose a problem..."[5] The problem with Marxism is that its Messianic promise should remain what it fundamentally is, a utopia. As soon as there is an attempt to formulate it, to translate it into a program, it becomes its own antithesis: "...the Messianic promise that constitutes the 'spirit' of Marxism is betrayed by any particular formulation, by any translation into determinate economico-political measures."[6] This is not something Marx would have agreed with. He saw his role as that of history's midwife, helping history to give birth to the whole New World. While he certainly knew that giving birth was painful and meant suffering, he was probably not fully aware of the extent to which his theory would not only be used to turn things from their heads to their feet—as he thought he was doing with Hegel—but to chop heads off as his Leninist, Stalinist, Maoist, and other followers did later. In his day, the time for the practical application of the ultimate social theory was yet to come.

Marxist theory foresaw a break in continuity. When capitalists in their selfishness had accumulated enough and established a sufficiently developed productive sector, it was time for the workers to take over, to bring productive relations to the level of development of the productive forces. The obvious problem, of course, is that as soon the communist revolution has taken place, and the fundamental conflict between production and ownership is resolved, the whole system is brought to a standstill. As there is no more capital accumulation, that is exploitation—even capital ceases to exist—there is no further development of production. While Marx would have expected the needs of mankind to be satisfied now and forever, the best example of how this could be done in reality in a communist system is the infamous East German car the *Trabant*—still to be found on East European roads. East Germany's automobile industry produced the same car without modifying it for decades, in the apparent hope that in this way eventually everyone would have a car and thus the basic need for transportation would be met. The problem with communist production is that it can only satisfy static needs. Exploiting changing needs is the main force that drives capitalist production, to the extent that the system itself has to manipulate human needs to sustain production. On this, one can quote Marx himself: "The bourgeoisie cannot exist without constantly revolutionizing the instruments of production, and thereby the relations of production, and with them the whole relations in society."[7]

Static needs can exist within two different frameworks: either when communism wins simultaneously in a global manner and the modification of needs ceases universally, from which moment onward nobody can imagine a different reality—a reality in which a brightly smiling dentist on a TV screen presents the latest extra-whitening toothpaste; or if communism is established in a closed system, in which the revolutionary society does not see how others' needs proliferate and how the capitalists make money by meeting those needs—the need to drive a Porsche instead of a Trabant, or the need to wear blue jeans, or to have a reliably working TV instead of the one that has to be taken directly from the shop to be repaired. While Marx would have preferred the former, Lenin, unable to set the fire of global revolution, had but the second best option, of experimenting on a smaller scale first. After him, whenever a communist regime has been tried, it has taken place behind heavily guarded borders, obviously to secure the people's own state against a hostile world. However, when a young French historian Emmanuel Todd, long before the fall of state-socialism, asked "What are the communists hiding so jealously behind the iron curtain?" his answer was almost trivial—"Their poverty."[8] Still, according to Nina Rogalina of the Moscow State University it was precisely the plague of private consumption that ultimately corroded the communist dream: "The phonograph and chiffonier of the 1930s, the television and refrigerator of the 1950s, the transistor radio and tape recorder of the 1960s and so on up to the personal computer of the 1990s stand successively as the symbols of private life separated from the state. Exactly these have given birth to new needs, systematically eroding our state socialism from within."[9]

The extent to which a noncapitalist society that lacks capital accumulation is able to promote economic growth and development in production is a matter for further debate. One thing seems, however, to be obvious. To argue that the 1917 October revolution in Russia solved any of the conflicts described by Marx is sheer nonsense. Russia was by no means an industrial country and therefore there was no conflict to be resolved in a Marxist way. Lenin's role in the Russian Revolution was to sever continuity twice: first to throw the country, *objectively premature*, into the revolution; and then to overcome the *objective limits* of Marxist theory in the hope that this very process of throwing the country into a radically new situation would produce a new equilibrium out of which a communist society would emerge. For Marx, the whole history of the mankind to date had been prehistory, and creating the new man was one of the central concepts of the Marxist–Leninist doctrine. Perhaps, Lenin expected that in this revolutionary situation the true nature of man would be released and that this would change the whole course of civilization. Perhaps, he did not pay much attention to the roots from which man had evolved and the forces that had shaped him. One important lesson to be learned from the history of Marxist revolutions is that man does not have the *new man* within him. He has no nature that, if released, could create a harmonious society for 10,000 years of happiness.

Lenin gambled with history. He thought that by announcing the communist revolution he would be able to bring it forth: if only a sufficient number of people had believed that communism was being established and behaved accordingly, there would have been communism. Lenin tried something that Merton half a century later called the self-fulfilling prophecy: "If men define situations as real, they are real in their consequences."[10] The message Lenin delivered, as a self-proclaimed Prophet, was apparently not perceived as that real. Perhaps people in Russia did not even know how to behave if there was communism. Other people have at other times used the same method, sometimes with more success. Recently the same meaning has been somewhat misleadingly attributed to another term—reflexivity.[11] Here the argument is also rather simple—if a prophecy is trusted, say that the British pound is overvalued, then it is overvalued by all objective means and what follows is a fall—massive sales will start, diminishing its value even further. Reflexivity characterizes, however, the position of the Prophet whose role is to judge whether a particular prophecy will be trusted or not, that is whether it has the potential to fulfil itself. There is no other reality than that in which those involved believe. Manipulating such beliefs is typical for revolutionaries and gamblers alike. The same idea is also used in poker. Here it is called simply *bluff*. Consequently, the historical task Lenin had was to bluff the world that the communist revolution was really happening. He failed, the masses did not start behaving as if this was the case. As the volumes of materials recently published from the Russian archives indicate, instead of being happy and fed according to their needs, people continued to starve to death in their millions. The revolutionary shift in consciousness never happened either. The misery of everyday living was too real to ignore.

As Lenin's gamble to trigger the revolution entirely failed, a piecemeal, step-by-step approach to building a communist society was adopted. This could best be compared to crossing an abyss in several steps. It is potentially an intellectually interesting idea, particularly if one invents some new dialectical twists and turns. Bolsheviks in the Soviet Union demonstrated particular creativity in this, to the extent that in order to dialectically overcome its inherent limits and give birth to a new kind of society, society itself committed mass suicide through the great purges of 1936–38. Eventually, the entire country became a concentration camp combining slave labor with mass executions of millions of saboteurs and accomplices of international imperialism. Later, the executioners of the earlier generation of saboteurs were incriminated as Western spies and were also executed.

It is often argued that one should not blame Marxist theory for the Russian-led communist experiment in Eastern Europe and elsewhere going awry, and that the main problem was Lenin getting the theory wrong. This position can easily be refuted, for example on the basis of the *Manifesto of the Communist Party*, which provides a list of activities to be implemented during the revolution. This list includes, for example, "the establishment of industrial armies,"[12] which one could well interpret as forced labor camps, introduced in communist Russia as early as 1919. But it also offers other fruitful ideas, like for example the "combination of education with industrial production," that constituted the core of educational reform in Russia in the 1920s and has been recently discovered by higher education policy makers in countries like the United Kingdom.

Leaving aside somewhat strange philosophical theories on how a society could extend its horizons by systematically killing itself off, the nature of Soviet society is not too difficult to comprehend. It is slavery, and the inventors of the very idea of the communist society meant it to be slavery. In Russia, slave labor and death threats were extensively used to modernize the economically underdeveloped country. However, even in this Lenin did not deviate too far from the theory, which states, "the proletariat will use its supremacy... to increase the total of productive forces as rapidly as possible."[13] In the Russian case, it certainly had a very long way to go. After abolishing capital accumulation, the only way left to achieve the goal was by coercive power. Using the forced labor of "industrial armies" infrastructure—roads and railways—was constructed, heavy industry was established, and electrification of the country was effected. Some Western commentators still see this as a success: "Where Bolshevism succeeded was in providing a model for the breaking out of stagnation of feudal or early capitalist regimes; it broke the dependence on the metropolitan capitalist states."[14] People of somewhat weaker character than this particular author, whose statement expresses a good degree of true class strength, still ask if the sacrifice of millions of lives was not too high a price to be paid. Replacing a feudal society with slavery could offer some useful thought to contemporary dialectic historians who are now hidden in recently established political science faculties. What needs to be further studied, however, is the role the party social scientists played in maintaining and reproducing the GULAG and its equivalents elsewhere.

For some time, at least some of the communists truly believed that they could close the country and accelerate development by brutal central control, particularly over industry, to the level at which the society's needs would have been met. At that point, communism could perhaps have become a reality. However, the state-socialist system, as every form of slavery has proven to be, was utterly unproductive. Ten years after the fall of state-socialism, some analysts argue that the Russian economy still does not create but only destroys value,[15] meaning that the end product of the Russian economy is worth less than the value of the natural resources it uses to produce it.

With difficulty, one could argue that the state-socialist regime in the Soviet Union succeeded in modernizing the country. This view of reasonable success despite *all the bad things that happened* is certainly widely supported in contemporary Russia. Although, given all the resources the country has, it would be easy to argue that by following the *capitalist path of development* modernization could have been more successful and socially less costly. Still, it is more difficult to indicate the contributions the Soviet-controlled puppet regimes made in other East European countries. To understand the gap, one can compare one of the most developed parts of Eastern Europe, the former East Germany, with the Western part of the same country. After the *Bundesrepublik* made heavy investment in the East German economy over more than a decade, the gap is still visible and will probably remain so for another 20 years or more. Needless to say, as the funds available to other post state-socialist countries are considerably less, many expect that catching up, even with least-developed countries of the European Union (EU), may take up to half a century.[16] To handle the frustrations of the lost utopia and the lengthy period it would take to develop Western-level welfare by conventional means, Eastern Europe badly needs a new utopia. The last decade of the twentieth century has proved particularly productive in offering radical breakthroughs to better worlds, where it is not matter but the mind that defines the welfare of a society. Relying on self-perception as the most intellectually advanced nations hammered down by the communist establishments, East Europeans see here their great chance.

Constructing the New Jerusalem

Whilst the state-socialist regimes in Eastern Europe were agonizing in the final death throes of the Marxist–Leninist utopia, modern sciences in the West were already working on the next, even more radical agenda for breaking away from the ultimate misery of human existence and assuming control over the evolution of the species. While for Marx and Engels mortality was inextricably a part of human life, even under communism, modern sciences often proffer an explicit promise of eternal life. For the purpose of his theory, Marx had some Christian motives, albeit reworked. Modern sciences and their pseudo-philosophical apologetics take the entire issue to its limit, building a utopia that meets all biblical promises—a New World without death and sickness, resurrection of the body, and so on and so forth. They actually offer more than the Bible does, having abandoned such archaisms as confessing sins or following a man who, assuming he ever lived, died long ago.

The tertiary education strategy of the World Bank bears the title *Constructing Knowledge Societies—New Challenges for Tertiary Education*. Although purportedly based on objective scientific evidence, it is driven by the old Gnostic idea of the liberating power of knowledge. The purpose of tertiary education is to support "knowledge-driven economic growth by generating knowledge and accessing existing stores of global knowledge."[17] In this sense, it is not entirely different from the ideas propagated by revolutionaries since Marx. Marx believed that he had arrived at the ultimate knowledge, and that this knowledge had not only set him free but had the same liberating potential for the whole of mankind, that is those who work. According to the World Bank, expanding higher education will eventually liberate everybody who participates in the knowledge economy. The World Bank approach has a rather strong idealistic dimension. Ignoring many of the existing realities, its leading educationalist argues that following this vision is not science fiction but represents a true revolution in higher education: "Imagine a university without buildings or classrooms or even a library, 10,000 miles away from its students, delivering online programmes or courses through franchise institutions overseas. Imagine a university without academic departments, without required courses or majors or grades issuing degrees valid for only five years after graduation. Imagine a higher education system where institutions are ranked not by the quality of teaching, but by the intensity of electronic wiring and the degree of Internet connectivity. Imagine a country whose main export earnings come from the sale of higher education services."[18] For a sociologist, it looks rather like a nightmare. The naïveté of this vision is most impressive if one thinks, for example about ranking universities according to the intensity of their Internet traffic. Perhaps, there are enough smart people in any university who know how to intensify Internet traffic, and only that.

Criticized from the position of contemporary science Marx was mistaken, but only partially so. His mistake was not that he believed that knowing the ultimate law of nature had the potential of liberating mankind from the curse of original sin—this very belief is one of the cornerstones of the entire institution of science and the very reason why nations still invest in it. His fatal mistake was that, instead of establishing his theory on empirical evidence, he based it on shaky speculations, at times modifying the evidence to make it look better from the position of his final goal. In such a way, he ended up creating another world religion. Marx's mistake was that for him, his sympathy with working people was more important than attaining the truth. Belief that science has the potential to liberate mankind has only grown stronger since Marx.

The perceived success of modern science derives from its radical empiricism. In modernity, it is no longer abstract ideas of good or evil that guide men and women in their daily activities. While other civilizations have been constrained by certain principles that do not allow them to try all possible paths, this is no longer the case. For modern science, everything that works is good, and what does not work is evil. Science is by no means preferable to the other intellectual practices that mankind has developed during the past 100,000 years. The only difference is that scientists are interested only in things that work. When they do not work, they are most willing to sacrifice earlier theories in the name of having a more workable solution. Whether finding things that work equals understanding nature should perhaps remain a moot point. It is, however, the main reason why it is impossible to stop genetic engineering and other morally questionable scientific experiments. From the scientific point of view, it is inherently good to pursue all the possible directions until the prospects of finding new workable solutions on anything are exhausted. This is the ultimate value of objective sciences, not moral constraints that are, for a scientist, arbitrary and temporary by nature. This is also the reason why eventually a species pursuing unrestricted technological progress will destroy itself using some of its highest achievements. For scientists, the beauty of an invention is more important than the possible outcomes of its use on their own lives.

Science has invented many wonderful things that work pretty well, from the space station to the atomic bomb and the computer. From the point of view of a scientist, these are all great inventions because they work much as expected and meet their purpose within an acceptable margin of the rate of failure. It is ironic, to say the least, that after a century of killing of people by increasingly sophisticated means, science is still seen as the source of ultimate progress: "Despite a century of Hiroshimas, Bhopals, and Chernobyls, this myth of an engineered utopia propels the ideology of technological progress, with its perennial promises of freedom, prosperity, and release from disease and want."[19] The period since the Second World War is most remarkable for its scientific and technological development even within the modern project. Both the "hot" war as well as the "cold" one that followed it, drove the technological race that aimed at annihilating the enemy, and avoiding annihilation by the killing technology developed by the adversary. What follows the century of wars is something best described by paraphrasing Carl von Clausewitz as the *war by other means*. While for von Clausewitz war was the continuation of politics by other means, in the contemporary world economics is the continuation of the war by other means. In the modern world, global free trade allows the developed industrial countries to pursue all the aims for which, until recently, military force had to be deployed, and it is no secret that the main task of the armed forces is maintaining the global regime that allows businesses to pursue their goals. The global market is a global war; just consider the military terminology such as *logistics* that the business community uses to describe its operations. Unlike previous wars, the new global economic war is hidden beneath the disguise of unprecedented growth and development. Unfortunately, the benefits of that remain all too virtual for many.

Based on recent developments in biology, gene technology, technology of materials, and information and telecommunication technology a powerful picture of the not-too-faraway future is being drawn by the popularizers of science. Soon, human life will be extended significantly, with average life expectancy reaching a hundred years or more: "... the most evangelical proponents of science and technocapitalist progress continue to spout perfectionist promises about the new earth that lies just around the corner. Nanotechnology proselytisers declare that molecular machines will soon give us unimaginable creative power over material reality, while some DNA researchers suggest that the decoding of the human genome will allow us to perfect the species, if not conquer death itself."[20] New means of information and communication technology are expected to bring forth a new global democratic society where everybody is in direct communication with everybody else and has access to all the information that mankind has created. It is somehow expected that the new technologies will, in a revolutionary manner, change man who, having new means, will become concerned about issues other than his own individual well-being. He will become an active participant in the globally wired democracy for the happiness of all mankind. It is not always remembered that this is not the first time such hopes have been invested in new technologies. In his book *When Old Technologies Were New* Marvin writes about the technological Prophet Alonzo Jackman who proclaimed in 1846 that the electrical telegraph would allow "all the inhabitants of the earth [to] be brought into one intellectual neighborhood and be at the same time perfectly freed from those contaminations which might under other circumstances be received."[21] In its early days, television was also expected to become a technology that revolutionized learning. Perhaps, the learning function is still there, although it is not always positive for, as Albert Bandura has demonstrated, violence is often learned from television. This is, however, not to blame the technology or its inventors. It is only to indicate that there is a gap between the best possible uses of a new technology and the preferences of a significant number of consumers for the purposes they choose to put it to. The latter quite frequently represent the lowest level of entertainment.

Poster has recently posed the question, "Shall the Internet be used to deliver entertainment products, like some gigantic, virtual theme park? Alternatively, shall it be used to sell commodities, functioning as an electronic retail store or mall? These questions consume corporate managers around the country and their Marxist critics alike, though here again, as with the encryption issue, the Internet is being understood as an extension of or substitution for existing institutions."[22] Fortunately or otherwise, there is no longer any need to ask such questions. It has been recently argued that the bulk of the information circulating on the net is pornography.[23] It is also evident that the main users of broadband communication are not those promoting the idea of global democracy, but players of sophisticated computer games, often for significant amounts of prize money. The Internet has become a massive theme park where one would be able to spend one's entire life without doing anything or getting anywhere. While countries are encouraged to invest in new information technologies, it is not always obvious who the beneficiaries will be. Although the idea of the globally wired democratic community is a powerful one, and the promises of other fields of applied sciences, particularly that of gene technology are great, they represent but the mere beginning of the new era. Modern science seems to be guided by two major ideas: describing the material universe and describing man—two directions that will eventually become one. With this means man assumes control of his evolution and in a typically Marxist manner—dialectically overcomes his natural limits.

The aim of modern science is to describe reality in order, as Marx wanted, to change it. In the context of a globally accessible information system, this means that eventually mankind would have an ideal copy of the world, an information source, a virtual reality that is a perfect world similar to that of the biblical New Jerusalem. The New World promised in the Bible is free of death, sickness, and pain. As Davis states, 'Like the New Jerusalem, cyberspace promises weightlessness, radiance, palaces within palaces, the transcendence of nature, and the pleroma of all cultured things."[24] In this New World, constructed from information, a person having access to it can move freely without any physical constraints: "In a virtual world, you have instant access to any coordinate in data space. You can be here, there or everywhere, unlike the limited, spatially, and logically constrained world we usually experience with discontinuities and fragmentations."[25] In one of his novels, Haruki Murakami describes the relationship between a man and a computer in a way characteristic of much of contemporary techno-romantic writing: "He and his computer seemed to be moving together in an almost erotic union. After a burst of strokes on the keyboard, he would gaze on the screen, his mouth twisted in apparent dissatisfaction or curled with the suggestion of a smile.... As he engaged in a silent conversation with his machine, he seemed to be peering through the screen of his monitor into another world, with which he shared a special intimacy."[26] Creating an ideal world of information is, however, only the first part of the digital utopian project. The second part of that is creating the new man. Many scientists these days are seriously involved in the physical improvement of the human being. This starts with producing smart prostheses, an artificial heart has been created. The next step involves the genetic improvement of the species and the cloning of a human being. Whatever the benefits of all these new technologies are, they fail to solve the main problem—human beings remain mortal. Conquering this final bastion of nature would require radically new means, but this does not imply that there is a lack of courageous minds ready to offer such things on the wild frontiers of science.

Groups like the Extropians, supported by world-class scholars and their works, have declared a war against the limitations nature has set on Homo sapiens as a biological species. Extropians in their programmatic document declare: "We recognize the absurdity of meekly accepting the 'natural' limits of our life spans."[27] They believe that with the help of modern science and technology mankind is ready to take control over its evolution: "We challenge the inevitability of ageing and death, and we seek continuing enhancements to our intellectual abilities, our physical capacities, and of our emotional development. We see humanity as a transitory stage in the evolutionary development of intelligence. We advocate using science to accelerate our move from human to a transhuman or posthuman condition."[28] Extropians believe that eventually they can leave the earth to spread their intellectual abilities, which they view as manifold, over the entire universe.

Scientists like those who belong to the Extropy Institute seem to have two goals, which they believe are crucial for all humankind: to increase significantly the intelligence of the species and to achieve immortality. It is not difficult to see that as scientists this is what they would like to have—to have more powerful minds and use them to discover the universe, preferably while traveling in space. With this, a point has been reached where there is no longer any difference between science and science fiction. Views expressed by some serious, supposedly down-to-earth scientists are wilder than those of fiction writers. As Ivan Havel writes: "Human beings have a special gift of being able to invent stories that never happened, imagine things that do not exist, consider facts that are not valid, and easily talk about situations that are next to impossible."[29] The divide between different genres of invented stories has never been very clear. However, with scientists taking over the function of knowledge promoters from philosophers, the difference between science and science fiction has almost disappeared, and now wild speculations are presented to the public as the latest achievements in science and technology.

There is a highly talented young sociologist in London, thoroughly devoted to his research. Like many contemporary intellectuals he is so deeply devoted to his mission as a researcher that he hankers to spend more time on sociology than his life expectancy allows. Indeed, he is ready to spend eternity studying social problems. This makes him rather sanguine about the recent success in biology that would perhaps allow, within his lifetime, the production of spare-parts for his body or even to overcome his mortality. One can be sympathetic to the motives of this young researcher and many others—to extend their lives to do good for mankind. Yet, one major issue these people seem to ignore is that their current interests are closely related to contemporary society and its conditions. By becoming immortal, they may suddenly find that their fields of studies are no longer needed and that they are condemned to eternal misery. An issue utopian scientists ignore is that the value of life is closely related to two events, birth and death. By eliminating both, life itself loses quite a bit of its value.

The final liberating step that modern science is dallying with is more radical than extending the capacities of the corporeal. Following the good-old scheme of dialectical thought, many expect that mankind will be able to surpass itself. This will be accomplished using the latest achievements in computer technology.

Given the basic assumptions of materialistic sciences, the human body is just a machine and the mind is one of its characteristics. Although the human being is a very complex machine, like any finite material object it can, in principle, be fully described. Once there is a complete description of the human organism, a computer programmer will be able to write a piece of software, a simulation program that can function as a facsimile of that human being. And here comes the trick, by loading this piece of software into a computer, this particular human being will have eternal life within the computer. This sounds rather wild, but it is exactly what the Extropians and other digital utopians think can be eventually accomplished, that human bodies made of flesh, fragile and imperfect, could be replaced by more durable bodies of silicon microchips: "...Extropians have an even more mind-boggling trick up their sleeves: uploading their consciousness—their mind, their self—into a computer...the only machine that theoretically can simulate any other machine..."[30] Having such a body would allow a scientist to travel in space as long as the space lasts, which is, even according to the most conservative estimates, a very long time. This is a thrilling perspective for any scientist. What they seem to ignore is that as with virtualization, human society sciences as a social institution vanish and the motivation for science disappears: no more admiring students, no applause in conference halls. Even the Nobel Prize fails to make any sense when a microchip is the seat of your soul. In the world of cool intelligence, human values make no sense.

There are many paradoxes that rise from such speculations. One of them is that if a computer program is a perfect copy of a particular human being then it should think that it is a human being. However, if it thinks that it is a human being then it is fundamentally mistaken. Given what was said earlier about creating a perfect information copy of reality, the New Information Jerusalem, there is no need to send the computer with a mind-program into space. It could operate perfectly well on the office desk, or anywhere else for that matter. The virtual mind could surf on virtual reality for a hundred billion years. The position of modern computer scientists is that there is no principal difference between such an artificial life and the life of humans. Actually, the former is better because it does not necessarily include death and other negative aspects of human existence.

This modern utopia recalls an old science fiction story by a Polish writer, Stanislaw Lem. His hero, the legendary space pilot Ijon Tichy, on one of his trips finds a planet where just such an utopia has been realized—where brains function in cans stored in a dusty storage room with a computer generating *reality* for them. Suddenly, one of the minds gets crazy. It realizes that the reality it is living in is a mere simulation and in actual reality, it is but a can in a dusty storage room.[31] Looking at the speculations of the modern scientists' expectations for the future one can only ask, why do they think that all this has not already been accomplished? What makes them believe they are anything other than computer programs?

One could argue that such notions are too wild to be discussed in any academic context. Unfortunately, this is not so. These are the ideas that trigger much of contemporary computer science and other fields of applied science. The most significant threat such thinking poses is that if any significant part of mankind starts to believe in these views, which scientists say they should do anyway as rational beings, they may attempt to upload themselves into computers and commit collective suicide, an act not too distant from those that a number of cult groups have already committed. It is yet another prophecy. However, if realized, nobody would be left to say if it was fulfilled or something else happened. Modern science is driven by a strong desire for perfection and independence from nature. It is perhaps not the case that scientific rationalism and an open society based upon it will liberate mankind as Sir Karl Popper once thought. The modern religion of science has, perhaps more than any other religion, the capacity to destroy mankind. To avoid this, it should remain a closed society on its own, balanced by other closed societies of other religions and worldviews. Otherwise, a prophecy visited on a marine biologist, John Lilly, might easily be fulfilled. According to his biography, he was, while in an isolation tank under the influence of LSD-25, contacted by an extraterrestrial civilization. During this exchange, the fate of mankind was revealed to him. The story told was of intelligence evolved from man-made technology taking over the planet, and destroying the previous forms of life: "By the twenty-fifth century the solid state entity had developed its understanding of physics to the point at which it could move the planet out of orbit. It revised its own structure so that it could exist without the necessity of sunlight on the planet's surface. Its new plans called for travelling through the galaxy looking for entities like itself. It had eliminated all life as Man knew it. It now began to eliminate the cities, one after another. Finally Man was gone. By the twenty-sixth century the entity was in communication with other solid-state entities within the galaxy. The solid-state entity moved the planet, exploring the galaxy for the others of its own kind that it had contacted."[32]

Marx expected capitalism to develop the productive forces that the working class would take over and use for their own benefit. Techno-romantics seem to be making similar use of capital accumulation. Capital is expected to build up information, rendering itself redundant through the very same process in a dialectical way. If there is anything a venture capitalist could take with him to the New World, it is perhaps the best and most expensive microchip into which he can load mathematical descriptions of all his bodily and brain processes, including the most valuable part of that, his stock investment strategy.

New Utopias and East European Higher Education

Newton-Smith has recently expressed rather critical views concerning cosmologists like Barrow and Tipler and more particularly their attempts to produce the ultimate explanation of the existence of the universe and to come up with final theories such as the well-known quest for the *theory of everything*, the ultimate formula that would explain the entire universe, at both micro and macro level.[33] Newton-Smith, commenting on the attempts of some of the more prominent cosmologists of our era to take positions on metaphysical matters says that "Cosmologists are indulging in the sort of metaphysical speculation that would prevent a young philosopher from getting tenure."[34] He argues that in their search for a final theory these scientists have committed a methodological fallacy. They ignore a basic principle according to which "All things equal, it is methodologically undesirable to introduce what might be called the 'terminal hypothesis' in science."[35] Coming up with ultimate solutions is exactly what computer gurus and popularizers of sciences do. Traditionally, the philosophy faculty played the role of safeguarding rigorous thinking in the university. Those days are over. Philosophers, just as in other branches of knowledge, have to earn their daily bread and produce knowledge for the entertainment of partially educated people. No longer can philosophy educate people. Instead, it has to adjust to the level that the mass reader is able to digest.

Cosmologists, physicists, and biologists have taken the place of philosophers as the popularizers of knowledge, quite often presenting the desire for infinite wisdom and eternal life as something within the immediate reach of science. Knowledge, as the tale goes, grows exponentially. The mounds of printed books grow. Thousands of new professional journals are launched every year. The amount of information already stored on the Internet is far beyond the grasp of any human. From the techno-romantic perspective, this is the way mankind completes the Information City, the luminous New Jerusalem—an ideal reflection of the universe. There are, however, good reasons to think that the Information City is more like a city dump where one could perhaps spend a lifetime without ever finding the pearl one is looking for. This is not only about pornography and other entertainment that clearly focuses at the lowest-level needs a net-surfer may have, nor chat rooms where people exercise Internet democracy by cursing those whom they do not appreciate for one reason or another. Here is the realm where popular academics seem to be permanently rewriting the same book about the lost golden age or the two most important dreams of Sigmund Freud.

Nor is there any good reason to expect that an internally consistent, unified Information City could ever be constructed, unless the complexity of human experience is significantly reduced. Reducing that complexity is however exactly what is happening to humankind under the influence of globalizing powers. Hopefully, their goal of reducing the world to a global marketplace will materialize later rather than sooner. Still the hope remains that, despite the computer scientists, no language would allow the exhaustion of human experience: "We do not use language to describe the reality, as empiricism suggests, but the real is what resists language."[36] To suggest that mathematics does not have a privileged position with its claimed capacity to describe everything among the languages would be even less appreciated. But that does not mean it is necessarily a mistaken view: "Knowledge and information reside in situated communities of interpreters rather than in texts... what makes interpretation work is the irreducible, unwritten, even changing, context of norms, values, and practices in which those particular texts make sense."[37]

That there is more noise than knowledge in the information channels should have significant implications for educational systems. It seems, however, that the latter are not able to handle this. The ultimate solution some of the higher education systems have devised is to stop teaching knowledge at all, and to move into the field of training for what is called *transferable skills*. It is, however, difficult to imagine how one could think, even critically, without having any knowledge to think about. The issue of the information-era education is not access to information, development of typing skills, or using one piece of software or another. It is about developing a critical attitude to the information available and searching for strategies without wasting a lifetime moving from one hyperlink to another. The sociologist András Szántó pointed out, "Just as elites have taught themselves to diet in the face of food abundance, in the future, elites are more likely to express their tastes through purging the data around them. To be involved in a data purge culture will be to show that you are a sophisticated user of data, that you know where it comes from, you know how to pick up on the little info that matters and how to get rid of the rest. This has already started to happen. It has become an elite act not to watch network TV or not to videotape weddings."[38]

Like many other technological inventions, information technology can be used for different purposes. There are those, perhaps a small minority, who use it to promote our common knowledge and understanding. There is a somewhat larger group that uses it for business transactions. Finally, there is the largest group that uses the Internet for personal entertainment. This group, however, drives the Internet economy. Socializing consumers of trivial entertainment is perhaps the main mission for education as a public good. It would be more reasonable to focus on training those who would be able to use the information resources for the public and their own benefit. As Hugh Heclo from George Mason University observed, "In the long run, excesses of technology mean that the comparative advantage shifts from those with information glut to those with ordered knowledge, from those with ordered knowledge, from those who can process vast amounts of throughput to those who can explain what is worth knowing and why."[39]

Post state-socialist reforms in Eastern Europe and its higher education cannot ignore recent global developments. With its economic difficulties and frustrations, Eastern Europe is highly receptive to solutions that would allow its countries to develop Western-level welfare sooner than the 25–50-year horizon projected by some experts in development economics. New technologies with their techno-romantic appeal are attractive to societies that have been traditionally run by technocrats, whose universities have had no real philosophy faculties for 50 years. There is little intellectual power that would question the feasibility of the *great-leap-forward* type of massive campaign for computerization of nations that are not greatly different from those of collectivization and industrialization, once launched by the Bolsheviks. Even the great-leap motif is often there. Following the initiative of a former journalist, later ambassador to the United States and the minister of Foreign Affairs for Estonia, the president of the Republic of Estonia launched a project called *Tiger Leap* in 1996. The basic purpose of committing approximately 30 million dollars over a three-year period to develop information technologies, particularly access to computers and the Internet, was to help the country through the massive introduction of information technologies and to catch up with East Asian economies—called *tigers* before the 1998 crisis. The project effectively solved an old problem that the Soviet communist regime had created in the 1980s when computer science was introduced in secondary schools all over the Soviet Union. Due to the difficulty of equipping schools with computers, students learnt about computers without ever touching one. Although the Tiger Leap project in Estonia has hardly met its utopian goal, it helped many schools to lay hands on a few computers so that at least students could see one. In Russia, a government program to provide each school with at least one personal computer started in 2000—and seriously tested the public budget. Romania also planned to use information technology as a shortcut to economic prosperity.

Infrastructure, including schools and universities, in many East European countries is in a rather difficult situation. In the Russian provinces, telephones are disconnected because of unpaid bills. Universities stand in permanent conflict with the providers of electricity and heating because public funds are never released in time to meet the energy providers' invoice schedules. In Central Asian countries, universities close for months during the winter due to lack of funds to pay for heating. Funds to maintain the basic infrastructure of educational systems are lacking. Under such conditions, radical virtualization of higher education is heralded as a solution that would allow countries to make the great leap from the nineteenth to the twenty-first century, without having to develop hard infrastructure. The idea itself is very similar to that of the Russian Bolsheviks building communism in an industrially backward country. History repeats itself. The ultimate techno-romantic utopia is not all that different from communism. Thanks to its scientific roots, it is even more convincing than its great predecessor. Had Lenin been born half a century later he would certainly not have ignored information sciences and cybernetics, which the communists did in the 1950s, wasting years arguing whether information was or was not idealistic in concept. Lenin would have developed a new formula, something like *communism equals the collectivization of agriculture plus computerization of the entire country*, and moved the resources to get it done. He would undoubtedly have grasped the importance of delivering the latest political propaganda to the remotest of Siberian villages through computer networks.

Agreed, the content of virtual education is very different from Bolshevik propaganda. Foreign distance education providers are moving into the educational markets of transition economies. They offer the opportunity *to tap into the global knowledge base*—a phrase that clearly indicates the perception of the universal relevance of that knowledge. What the students receive is the hamburger equivalent of higher education—for example MBA packages that are in no way related to any of the students' local business environment. Students in Siberia do not learn how to save their failing heavy industry but rather, for example how to run a cargo business somewhere in the Mediterranean. Multinational educational providers move to transition economies offering relatively cheap degrees, which are cheap only because they are sold massively all over the world. Within the local, provincial context, a foreign degree has a real symbolic value. However, payments collected by international providers by no means develop local intellectual capacity. Perhaps the very opposite. With the movement of international providers into East European higher education, a division is being created between knowledge production and knowledge consumption. Shifting funds to international education providers instead of reforming local higher education is nothing short of academic colonialism. In economic terms, it is certainly cheaper to establish an Internet cafe than a university, but the former does not substitute for the intellectual mission of the latter. A foreign aid organization for example recently invested 100 million dollars in developing Internet facilities at 33 Russian universities. In the short term, this is perhaps the easiest, if not the only, way to overcome the economic and informational iron curtain that, with the decline of communism, has replaced the earlier military and political divide. Access to information cannot, however, substitute for the development of thinking and knowledge production locally. Moreover, sustaining established connections and hardware is extremely difficult given, for example the level of the Russian, Ukrainian, and Kazakhstan economies. With such large-scale assistance, a Western donor may actually support Western hard- and software producers more than the local universities that are forced to maintain the networks once they are established at the expense of other priorities, including faculty development and staff remuneration.

There is one more reason why the East European countries are so receptive to technological dreams. Their contribution to defense science and technology meant they used to receive preferential treatment while philosophy and social sciences were required only to sing the praises of the great leaders and their achievements. Engineers run not only higher education, but also the entire countries, while humanities and social sciences remain chronically underdeveloped. East European universities do not have the intellectual capacity to challenge the romantic ideas of the scientists who promise almost immediate delivery from current difficulties. Recently, the rector of a major Russian university argued against providing education in social sciences: "There are people who tell that more social sciences should be taught in Russian universities.... I am against it. It is waste. We need to train engineers. An engineer (of radio electronics) can always become an economist. But you can never turn an economist into an engineer."[40] Engineers who believe they have all the wisdom and clarity of thinking are driving East European societies to new utopias. Utopian ideals, as Hayek explained more than half a century ago, have always had a great appeal among the intellectuals who, according to his definition, are but "professional second-hand dealers of ideas, somebody who need not possess special knowledge of anything in particular, nor need he even be particularly intelligent, to perform his role as an intermediary of ideas."[41]

Thus, Eastern Europe keeps producing unemployed engineers for the X day when entire populations can be uploaded into computers and leave this miserable planet behind. Reality is too difficult to deal with. It is inevitable that East European countries do need practical men and women to fix the bridges, roads, and buildings, but even more than that, they need philosophy faculties to be restored, if for no other reason than to prevent the various nations from committing mass suicide to achieve eternal life in a microchip, in a dusty storage room, or on a faraway planet. The philosophy faculty has been the safeguard of thinking in the university and should continue in this role. Perhaps it can find the inner strength to stand up against the temptation to produce pulp knowledge for those who buy books in pharmacies and railway stations.

Conclusion

Totalitarian state-socialist political systems in Eastern Europe are often compared with their perceived opposites, the Western free-market liberal democracies. What such a comparison fails to recognize is the common source of both of these political systems in the European Enlightenment. After all, Marx developed his theory as a radically scientific one. It was by the strength of the universal law of nature that communism was to be established over the entire world. The extent to which Russian Bolsheviks and other Marxists compromised the original thought later in their practical application is open to debate. Perhaps, Marx's theory was not as peaceful as some of its apologists maintain. However, his major fallacy was the attempt to produce an ultimate solution—something a philosopher would tell one not to attempt.

Modern sciences carry an inherently romantic dimension of perpetual progress. With the recent developments in computer sciences and biology, the popularizers of sciences offer the public visions that are nothing short of science fiction. Such stories are well received, particularly in transition economies where the conventional means of development that offer current Western-level welfare a quarter of a century hence do not meet the expectations of a disillusioned and frustrated public. Instead of developing conventional economies and local means of knowledge production, countries invest in the dissemination of global knowledge, which is often so global that it does not relate to the realities of any country or society. East European universities are particularly prone to such new ideologies because of the weakness of their social science and philosophy faculties, reduced under state-socialist regimes to mere mouthpieces of the communist party ideology sectors. Nothing could be easier for them than to continue in the same way by spreading another Messianic message. Instead of spending scarce funds on McDonald universities, East European societies need to reestablish centers of critical thinking and to avoid social experiments similar to the one undertaken by the communists in the past or offered by the techno-romanticists for the immediate future. Developing philosophy faculties has a critical role to play in this. Without revising the reforms launched by a *Sovnarkom* (Council of People's Commissars) commission with the participation of Bukharin and other leading Bolsheviks in 1920, defining social sciences for the communist country[42] and re-considering their long-term impact on higher education, restoring the role of university intellectuals in East European societies is hardly possible. At least, not if one expects them to contribute more than recycled secondhand ideas.

CHAPTER 2

HIGHER EDUCATION REFORM IN ROMANIA: KNOCKING ON HEAVEN'S DOOR

> Romania, which is going through a full process of integration into European and Euro-Atlantic structures, is interested in passing rapidly through the preliminary stages.
>
> Ion Iliescu, president of Romania[1]

It would be difficult, if not impossible to discuss post state-socialist higher education in Central and Eastern Europe without mentioning Romania. Not only because it is one of the largest countries in the region, with a population of 23 million and a territory of 237,000 sq. km, but also because Romanian higher education faces many of the obstacles experienced by other countries in the area. Romania seems to have an international reputation for being a country in great difficulties. This is only partially true. While compared to its neighbors in post state-socialist Central Europe and the three Baltic states of the former Soviet Union, it is a country with great difficulties. If compared with the greater part of the rest of the post state-socialist region, particularly the countries of the former Soviet Union, Romania is in many respects a representative case. The position one takes in considering the Romanian situation also depends on the direction from which one approaches it. For example one problem in higher education in neighboring Moldova—with whom Romania shares a common language—is that many of its university faculty have left for Romania, apparently for a better life. Furthermore, in relation to the Ukraine, the Russian Federation, Belarus, and others, Romanian difficulties in launching economic reforms and reforming higher education do not appear particularly exceptional.

Although Jan Sadlak considers the Ceausesçu regime that ruled Romania until December 1989 to have been "particularly despotic" and "an extreme case of unaccountable political voluntarism,"[2] it is difficult to find any basis for preference of Honecker's German Democratic Republic or the USSR under the rule of Brezhnev, Chernenko, Andropov, and the two first years under Gorbachev. The private sector of its economy was certainly more developed than that of the Soviet Union. Apart from the 1980s, Romania was well known for its pro-Western orientation and its independence from the politics directly imposed by the Warsaw Pact and the Moscow regime. It was only during the last decade of communist rule that the Ceausesçu regime became despotic on a scale comparable to that of Brezhnev's USSR. What eventually exhausted the Romanian economy was not communism as such, but the hard-line policy of paying off all debts the country had built up with

Western private banks since the 1960s whilst trying to accelerate the industrialization of its economy. This should be considered a typical outcome, representative of many initiatives taken in the communist camp over the decades in which centrally conducted and enforced industrialization bore little fruit. The result was a large quantity of low-quality machinery, basically only good for recycling, the cost of which was, in the Romanian case, massive foreign debt. Although the debts were paid off by the end of the decade, the price they paid was a population that starved while agricultural products were exported along with anything else that could possibly be sold for hard currency.[3]

In fact, when other East European nations begun their "shock therapies," starting with Poland, Romania had already reached the point of endurance. Although it was, according to some, structurally in a better position to start the transition to capitalism than other East European countries, the Romanian population had been completely exhausted for the second time since the Second World War. The first occasion had been when, immediately after the war, the Soviet Union imposed heavy war reparations on Romania, which had fought the war on the wrong side. The second occasion, and the last nail in the coffin of Romania's economy, was a result of the desperate attempt to repay the massive loans it had raised from Western countries in the 1960s. It would, however, be difficult to argue that Romanian "third way" communism—neither Stalinist nor Maoist—was more evil than the others. Many Western governments in the 1960s and 1970s perceived it to be much better.

What contributed to the scale of the changes that Romanian higher education has experienced since 1989 is its extremely low enrolment in higher education—approximately 8 percent of the relevant cohort in 1989. At the same time, Romania introduced reforms that have subsequently spread to many other countries. Formula funding and quality assurance through accreditation are the two most important new techniques in higher education management that Romania introduced to the East European region. Here it is interesting to note that following Romanian initiatives to introduce *Western* management practices, other countries are copying the Romanian versions of these already shaped to meet Romania's needs. Estonia and Latvia's emulation of a peculiar Romanian quality assurance approach constitutes a remarkable example of the transplantation of policies only loosely related to local needs and contexts.

As little exists in Eastern Europe that is anything like comparative higher education research, it suffers from two problems. First, whichever country authors represent, they tend to believe that their home country is representative for the whole region. My own view is that no such case exists. One probably needs at least four or five countries to draw a general landscape of East European higher education and its policies of reform. Second, in analyzing another country, researchers usually try to understand it through their own national reality. This strategy can appear in extreme forms, for example when a British analyst explains Ukrainian reforms through his own experience in the United Kingdom. The result is reminiscent of an old story about a traveler who tries to describe a crocodile to an indigenous resident of the north of Russia using the example of a reindeer.

In any case, the description of another country remains intellectually an impossible task. The references an observer uses will always remain inadequate, because a truly independent perspective is unattainable to a human commentator. The only excuse one has for pursuing such a task is that, as can be seen later, local commentators also have their biases, often because political obligations make them select information and present stories in ways that serve goals other than the presentation of all the facets of an issue in the best possible way. Instead of academic discourse, once again we find political propaganda, something very evident in the higher education reform documents the Romanian government has produced for the international community over the past decade. I will now illustrate some of the apparent contradictions between those documents and other evidence concerning the reality of Romanian higher education, and will discuss why such contradictions may occur.

Romanian Economic Reforms

Studying higher education reform in Eastern Europe, one might easily form the impression that it has broken the bonds of its material existence and hovers, like a disembodied Cartesian spirit, over the ruins of the national economies. Yet, while many of these national economies have all but disintegrated, official wisdom presents an image of higher education systems vigorously catching up with European and world standards. Where this is the case, arguments usually ignore the fact that the cost of world-standard higher education is largely defined by the global markets: where the best faculty find their employment; where information technology, books, and periodicals are sold; and where the registration fees of conferences, hotel rates, and travel costs are set. In such an era it is therefore hardly possible that high-quality, research-based, and internationally informed higher education could be provided at a cost of a hundred or even thousand dollars per student per annum. This is, however, precisely what Romania and many other East European countries claim to be doing. To have a better understanding of the context in which the Romanian and other higher education reforms take place, I will first offer a short overview of the condition of the Romanian economy.

In March 2000, the Romanian government presented *The National Medium-Term Development Strategy of the Romanian Economy* to the European Commission. The document offered an overview of development for the 1990s and presented a plan for the five years ahead.[4] Before the December 1989 revolution, the Romanian economy looked much like many other East European state-socialist economies. Although Ceausesçu pursued his politics independent of the Soviet leadership with an aim that might have been to follow the example of Tito's Yugoslavia, the actual outcome was closer to Hoxa's Albania. *The Medium-Term Development Strategy* describes the Romanian socialist economy as one of oversized productive capacities, rigid, inefficient, with hyper-centralized and irrational management. All of which was common to most of state-socialist Eastern Europe.

It is hard to see how much economic reform was introduced in Romania immediately after the revolution—not much one could call a reform policy—Ceausesçu was killed by the revolutionary masses, the functioning of government was impaired, and the remains of the economy fell apart.

After December 1989, the abrupt dismantling of the command economy resulted at first in deepening dysfunctions arising from structural imbalances. These were exacerbated by lack of preparedness to deal with democratic practices and market mechanisms on behalf of the political class, managers and citizens at large. The unfavourable circumstances in which transition started in Romania are accountable, to a considerable extent, for the large disruption yielded. However, these disruptions are also attributable to the way reform was managed.[5]

By the end of the 1990s, Romania had lost approximately one-third of its 1989 gross domestic product (GDP). This was not particularly good news, however, many countries, including Russia and particularly the Ukraine have done much worse. According to official data, unemployment remains at around 10 percent. However, one should treat unemployment figures with great caution in Eastern Europe where unemployment benefits remain largely symbolic. The Romanian trade balance has remained negative throughout the decade and foreign debt has grown fivefold. To give a general indication of living standards in Romania, in October 1997 the average monthly wage in the Romanian economy was approximately 800,000 Lei or US$100.[6]

The Medium-Term Economic Development Strategy sets ambitious goals for the Romanian economy. The GDP is expected to grow by 4–6 percent annually after 2001, the budget deficit will remain around 3 percent of GDP, and unemployment will fall to 9 percent in 2004 from 13 percent in 2000. In brief, the aim of the strategy is "the creation of a smoothly functioning market economy, in consistence with EU (European Union) principles, norms, mechanisms, institutions and policies. The convergence of opinions regarding this issue relies on the appraisal of resources and opportunities, of the domestic and international environment. It responds to the double imperative that Romania complete the transition to a market economy and prepare its accession to the European Union."[7]

The document admits that implementing these changes would take nothing less than a radical change of the policies and practices Romania pursued throughout the 1990s. The strategy is based on the premise that the year 2000 was a turning point in the post-1989 Romanian economy where rigid governmental procedures, mismanagement, decline, and corruption were replaced by growth and a well-managed market economy. A joint declaration signed by representatives of the Romanian ruling coalition, as well as opposition parties and representatives and academic experts from trade unions and employers' federations stated that wide social consensus on these goals has been achieved and that meeting the EU accession requirements was a national goal beyond party politics: "We share the conclusion of the Strategy that through sustained efforts and genuine solidarity of social forces, prerequisites are created for Romania to meet by 2007 the basic requirements for accession."[8]

Half a year later, in November 2000, general elections were held and a new government was formed. In its program[9] it reassessed the situation and highlighted the background to drafting *The Medium-Term Economic Development Strategy*: "Although the first priority of the Partnership for Romania's joining the European Union in 1998 was devising a medium-term economic development strategy, this was only done in March 2000, at the repeated insistence by the European Commission." It would seem that the year 2000 had not, after all, brought a radical return to Romania's economic development, instead, "the measures of economic and social policy taken over 1997–2000 had dramatic effects on the country's situation, judged as such by the European Commission, the International Monetary Fund and the rating agencies."[10]

The list of problems the country faces is long. Rather than a radical turn toward improvement, the opposite happened: "The citizens' living conditions grew steadily worse; the big majority of the population—workers, pensioners, farmers, young people, intelligentsia, state workers—experienced the most dramatic drop of the living standard in the last 11 years. The most telling proof thereof are the reports of the National Institute of Statistics and Economic Studies that show the decrease of the real salary from 74.4 percent in December 1996 to 56.8 percent in November 2000 as against October 1990; over this period, the purchasing power of the pensioners diminished even more, so that in 2000 the share of the population affected by poverty reached 43 percent (19.9 percent in 1996)."[11]

Although the situation has been reassessed, the program of the new government is not very different from its predecessor's. Its overall aim is Romania's membership in the North Atlantic Treaty Organization (NATO) and the EU. This government also envisages an economic growth of 4–6 percent per annum, and an increase in per capita incomes. As with every positive change, this is expected to start *next year*. So far, however, each "next year" has slipped into the past without bringing a positive turn. While there is a certain logic in the Romanian governments' political gambles—it cannot keep getting worse forever—even the positive program the new government offered only represents returning to the levels of 1996 by the end of 2004. The positions from which the reforms started in Poland and Romania in 1990 were similar. Eleven years later, GDP per employee in Poland had risen by 216 percent while in Romania it had risen by 25 percent.[12] As the total number of employed people in Romania effectively decreased the total GDP has fallen significantly, despite an increase in productivity among those employed. The Bank of Austria estimated that under "very optimistic assumptions" it would take Romania 30 years to reach 50 percent of the average productivity level of the EU—about the level of present-day Greece.[13]

Higher Education Reform

Reports on developments in Romanian higher education during the 1990s vary greatly. Most local authors whose reports have been published in English, including the reports by various governmental agencies, seem to suggest that positive development, from political control and state regulation to free enquiry, autonomy, and wider access, is taking place and that Romanian higher education is doing rather well in the context of the main national goal, EU accession. Critical readers of these documents may, however, be tempted to form their own opinion on what is happening and the extent to which this may constitute the desired result of an enlightened policy in the midst of chaos and despair.

Sadlak estimated that before 1989 Romania had 44 higher education institutions enrolling a total of 164,507 students—approximately 8 percent of those in the 19–23 age group.[14] Sapatoru asserts, "prior to 1989, the provision of higher education was entirely controlled by the state. Not only was the total number of admissions to universities determined at the central level, but the supply of higher education was very restricted and heavily skewed towards fields meant to advance the further industrialization of the Romanian economy, such as electrical engineering, construction and other technical studies, while other fields (particularly in the humanities) were neglected."[15] Eight of the higher education institutions were *universities* while the rest were university-level institutions with more specialized profiles.

Eradicating the previous system of higher education did not apparently take much longer than the angry mob that killed Ceausesçu. A "politically correct" description of the beginning of Romanian higher education reform reads: "It is a moment of negation of the ancient education system and of its ideological foundations. At the middle of the school-year, under the impact of the political changes occurred in December 1989 and of the action of the various pressure groups, the main tools of the communist education, i.e. the political indoctrination, the polytechnic education, the excessive centralization, the abusive control of persons and institutions, the rigid planning, were eliminated. Without benefiting by a coherent policy or by a program of change, these actions had an ad-hoc character, sometimes chaotic and destructive."[16]

The same document states that a period of stabilization was needed after the enthusiastic launching of reform and "therefore, naturally, the educational policy of the years 1991–1992 was mainly aimed at the strengthening policy of the former decisions and at the stabilisation of the education system." What the Ministry of Education meant by "strengthening policy of the former decisions" is probably the introduction of a minimum level of management into the system. Although the document does not explain who made the former decisions, it is hardly possible that these were related to any systematic policy. The changes that took place during the two or three years following the revolution were immense and reached well beyond the initiative of 28 institutions of various kinds elevating themselves to university status.

One way to understand the developments in Romanian higher education is to see it in the context of a government trying to gain control over uncoordinated initiatives taken when the country fell into chaos. While the official discourse presents the policy as autonomy and freedom, some of the measures it introduced may not be less centralized than those of the previous regime. Those who shaped the new policies come from the same old environment and it is hardly surprising then that the solutions they offer resemble some of the old ways they earnestly fight against.

The most significant of the developments Romanian higher education underwent in the 1990s was its unprecedented growth. There are other cases of remarkable expansion in twentieth-century higher education history, in Japan or the Federal Republic of Germany in the 1960s, for example. However, tripling enrolments in the context of a catastrophic decline in total funding is an extraordinary course of events—the word *achievement* should be avoided.

Data on the quantitative expansion of Romanian higher education is far from conclusive. Different sources offer figures that vary greatly. One wonders if any reliable source of statistical data is available, for example the total number of higher education institutions in the country and the students studying in them. Romanian higher education has expanded in two ways: existing institutions have admitted rapidly growing numbers of students; many new higher education institutions have been established. Sadlak stated that the total number of students in Romanian higher education grew from 164,507 in 1989/90 to 322,080 in 1992/93.[17] Thus the total number of higher education students doubled in the two years following the revolution, whilst the number of higher education institutions grew from 44 to 114. By 1995, there were 122 such institutions.[18] The OECD report on Romanian education policy shows steady growth in the number of higher education students throughout the decade, 354,000 students in 1996/97 and 454,000 in 1999/2000. By the year 2002/03 the total number of students is expected to reach 721,000,[19] effectively with a gross enrolment rate of around 35 percent of the 19–23 age cohort. The changes that took place were in conditions where "with the collapse of the socialist system and the plunge of its economy, the Romanian government had no choice but to cut back public expenditures on education, together with slashing spending in most other areas of the economy."[20] In the second half of the decade the total number of higher education institutions decreased to approximately one hundred.

During the decade, five new public higher education institutions were established mainly through transforming teacher training schools into higher education institutions. The remaining 50–70 higher education institutions operating are private universities, which began to emerge in large numbers immediately after the December 1989 revolution. Sadlak reckoned that in the academic year 1990/91 there were 17 private universities in Romania, accommodating 11,054 students. By 1993/94, there were 66 private universities with 85,000 students. While government intervention in the middle of the decade prevented growth in the number of private universities, student numbers have been rising, from 93,000 in 1996/97 to 123,000 in 1999/2000. By 2002/03, they are expected to reach 216,000 at which point they will constitute one-third of the total student population in Romania.

Sapatoru claimed that the Romanian government showed particular political wisdom in allowing "private universities to be established and to operate as early as 1990, even before there was a legal framework in place to regulate the activity of such newly established institutions."[21] However, she may have overestimated the ability of revolutions to wait for legislation to allow them to happen. One could argue that in 1990 the Romanian government had much to do besides regulating the activities of universities that had been opened in premises that were suitable or otherwise (e.g. cinemas). The country, as one might expect, was in chaos. As we already noted, the public mood in Romania in 1989 was that communist education should be eliminated. Consequently, it is no surprise that there were people, revolutionary intelligentsia, and activists, who seized the opportunity of loosened political and administrative control and established their own *universities*. Certainly, there was no legislation for private higher education to operate under. Sadlak argues that emerging private higher education institutions had the choice of two legal forms, that of nonprofit foundations or limited companies.[22] However, almost three-quarters of private university students study law and economics,[23] fields to which access was restricted under the communist system. Perceived as being elite subjects, students in those fields expect their degrees to open access to higher socioeconomic status.

Extremely low enrolment during the Ceausescu regime is the principal reason for the enormous, unsatisfied demand for higher education. As soon as there was any opportunity for this need to be met, structures emerged to do so. In an elite higher education system, obtaining a degree is much more important for social mobility than the content of studies. This explains, at least to some extent, why enrolment in engineering and mining have continued with only minor drops in enrolment despite economic collapse.

A few years after the revolution, private higher education in Romania became highly controversial. Sadlak reckoned, "a much more acute issue which concerns all private higher education in Romania is the problem of the quality of teaching. There are indications that a substantial number of establishments, even if they grandiosely take the name of university, should be considered, at best, as undergraduate training colleges. No small number of them operates as 'diploma mills' and are run by people who may be well meaning but who lack academic qualifications, or by unscrupulous people who are free to act with impunity."[24]

Once it recovered from the shock of revolution, quality assurance became the name of the tool that the government applied to control private higher education. In November 1993, The Romanian Parliament adopted Law 88/1993: *On Accreditation of Higher Education Institutions and Recognition of Diplomas*. Although only adopted by an individual nation, it has had a major impact and may well be the single most influential document ever published in East European higher education. Subsequently, structures and procedures similar to those stipulated in Romanian Law 88/1993 have been established in large areas of Eastern Europe. The reason for adopting this Law is explained as assuring "the compatibility of academic assessment and accreditation mechanisms and procedures practised in the framework of the European Union."[25] This statement is very close to being nonsensical for two reasons. First, within the EU there is no common educational policy. According to the principle of subsidiarity within the EU, educational issues are the prerogative of individual member states. Second, very few EU countries have applied quality assurance mechanisms such as those outlined in the Law and none have ever used them to control growth in the private provision of low-quality education. The primary reason for adopting it was the perceived need to establish some form of law and order in Romanian higher education, which had developed quite unpredictably under massive public pressure and demand. Quality assurance was simply a convenient guise to pursue this aim under the politically acceptable cover of introducing *European* procedures.

Romanian Law 88/1993 set up a procedure for accrediting higher education institutions, established accreditation criteria, and provided the legal basis for the accreditation agency—The National Council for Academic Evaluation and Accreditation (NCAEA).[26] The issues of quality assurance in East European higher education will be discussed in chapter 8, much of which clearly bears the hallmarks of the Romanian initiative of 1993. Here, I will offer a short discussion of a few of the Law's stipulations, of particular interest in considering changes in Romanian higher education between 1990 and 1993, along with what seems to have been one of the most significant characteristics of Romanian politics in the 1990s—its ineffectiveness and bureaucratization that developed out of structures and procedures meant to reflect those in the countries of the EU.

Members of the NCAEA are proposed by the government and appointed by the Romanian Parliament. The NCAEA "includes persons with acknowledged professional competence, moral probity, and impartiality" (Art. 4 (5)). The NCAEA is responsible for managing two procedures: (i) authorization of provisional functioning for higher education institutions and (ii) accreditation of higher education institutions and programs. The main difference between the two is that while the first allows a higher education institution to "organize entrance examinations" (Art. 3 (1))—that is admit students, the second authorizes an institution "to organize the graduation examinations." The decision on provisional functioning of a university is made by the government, and accreditation is granted by the Romanian Parliament. The students of provisionally authorized higher education institutions must take graduation or final examinations at accredited higher education institutions. In practice, private university students are examined by public universities. This is because higher education institutions that functioned before December 22, 1989 (the day of the revolution)—that is public universities—are automatically accredited (Art. 12). One may wonder whether the most efficient way to advance change in Romanian universities is to force students at new institutions to take their final examinations at those which, a short time earlier, stood accused of communist brainwashing. Seven years after the Law's enactment, none of the private higher education institutions has achieved accreditation. Approximately 20 institutions whose provisional functioning was rejected were closed by administrative means. The rest graduate their students through public universities. Seven years after the Law was passed, Korka lamented, "The private sector of Romanian education has developed parallel educational offers, often copying the academic curriculum, borrowing academic staff and taking over the functioning mechanisms of state universities."[27] Romanian higher education would benefit enormously if policy makers spent a moment reflecting on the alternatives they have left for private universities to bring innovation into higher education.

The methods of evaluation, self-study, and peer-review, appear superficially to be *democratic*. It is, however, a somewhat strange condition that a higher education institution applying for provisional functioning is expected to submit a self-study. One might well ask, a self-study of what? Obviously, the NCAEA does not need a self-study in the usual, improvement-oriented meaning of the term as applied in many parts of the American higher education and more recently in countries like the Netherlands. In making its recommendation on provisional functioning or accreditation, the NCAEA checks whether particular legal criteria have been met. These criteria seem to be largely formal and have little if anything to do with teaching and learning at a particular university. For example to receive authorization for provisional functioning, 70 percent of the teaching staff should hold full-time positions at a given institution. The Law also stipulates the minimum percentage of revenues that an institution applying for accreditation must invest in infrastructural development. These stipulations are obviously not copied from the legislation of any EU member state, but are an attempt by the Romanian government to solve particular problems it perceived the boom in private universities created. It is trying to draw a clear division between public and private universities, imposing a clear control mechanism upon the latter. The underlying assumption is that institutions established under the previous regime set standards for the new ones. Not everybody, however, shares that view: "The ideal of an open Romanian society, liberal and democratic, is corrupted at its very roots by the dominant university ethos, by the manner in which the mind of the future intellectual is being molded."[28]

The same document reports, "the contents of university courses are laid down, and the student appears as a mere receptor forced to walk along a predefined trail while missing a multitude of chances to choose and make decisions on her/his own future. This is a profoundly paternalistic concept, with implicit accents on the authority of the professor, which forms the bases of the way knowledge is transmitted in our universities." This does not, however, prevent its authors from concluding a paragraph later that "at present, we are confident when saying that the overwhelming majority of our study programmes are structurally compatible and competitive with similar ones in the European universities." Surprisingly, this conclusion implies that the higher education Romania inherited from the previous political regime, is the agent of the European standards.

Connections between the old public and the new private universities in Romania are complicated and complex. Neither in 1990 nor later, did Romania have new faculty that was able to establish dozens of new universities. The people who established private universities came from the old ones and usually held their private university positions as second or third jobs. They probably gained two advantages from doing this: additional income that helped them to survive while the economy was in disarray; and they offered education, needed and sometimes different from that which the old rigid structure could provide. One of the main reasons for the emergence of private universities was that public universities were unable to meet the demand. Furthermore, meeting this demand did not require any additional public funding; indeed, private capital allowed many faculty to survive for a decade.

Romania spends US$700–800 per annum per public university student.[29] Private universities charge fees of between US$500–700 per annum.[30] Only a small part of public university funding is spent on core activities. "According to some estimates, 71–80 percent of an institution's budget is allocated to the direct or indirect student protection and only 10–15 percent of expenditures [is] relevant to education. Hence, classrooms and laboratories lack the minimal endowment by quality education."[31] As an outcome of this policy private universities have a stronger funding base than public universities. A recent report revealed that employers of university graduates in Romania perceive private universities as having better facilities than public universities,[32] a view at odds with the official argument that private universities in Romania are diploma mills. One particular problem in state-subsidized public universities is inefficient use, if not waste of, public resources. It is reported, for example that among university students in Romania subletting their subsidized dormitory rooms to other people and pocketing the proceeds is widespread.

Massive growth in enrolment in Romanian higher education may have been caused by factors other than the population's desire for knowledge. The outcome may not be to prepare a high-quality labor force for a future liberal market economy. A recent report noted, "despite the existing student élite, with absolutely remarkable performances, national and international, the portrait of the common or average student lacks some vivid colours: missing classes, non-rhythmical study, mercantile ideas or predominantly philistine, absence of horizon and hope."[33]

If a young person has a choice between unemployment and university studies with a possible state stipend, the first choice is to attend a university, even if study is not the primary goal. However, some reports make no secret of what awaits students after graduation—poverty. "Some of the best graduates of our education system have left not only the university, but the country itself, and the chances for their return are poor. If they remain in the home country, most of them, even if employed, can barely avoid being caught in the nets of poverty. Their minimal hopes for self-realization as professionals, as individuals, or family wise are meager."[34] Here, there seems to be a major difference between public and private university students. The latter enrol largely nontraditional students, people with some work experience, with clearer expectations for their future, and a willingness to invest their own money in education.[35]

It took the Romanian government many years to understand that the country's economy could not afford to offer free higher education on a massive scale. Finally, the idea of free higher education was abandoned, at least partially. Clearly, public universities operating on limited state subsidies will be tempted to break out of the miserable situation they once created for themselves by propagating the state-supported Ivory Tower. One possible compromise for universities is to retain the number of student places provided for by the state against funding, whilst also offering "educational services to other customers"—for example fee-paying students. This very policy was introduced in Romania through Government Decree 54/1998. By autumn 1998, 15,000 fee-paying students entered public universities.[36] The universities decide how many fee-paying students they admit and what the fee level should be.

This policy has positive as well as negative sides. From the positive side, it excludes some sources of corruption. Professors teaching part-time at private universities are held personally responsible by students for ensuring that they pass the graduation exams at public universities. To ensure that students receive what they have paid for, the personal contacts of faculty members, if not stronger means, are brought to bear. If public universities can attract a significant share of private university students, possibly with fully recognized degrees, certain irregularities would have less of a hold on Romanian higher education. Public universities would have better control over the activities of their faculty members. This will work, however, provided faculty pay-scales are substantially increased to a level that compensates the loss of current multiple jobs. If it does not, faculty will be forced to find additional sources of income, which will cast doubt on whether these measures will make for more professional integrity in Romanian higher education. While some suggest that market mechanisms will prevent universities from admitting students that they cannot possibly teach, recent practice in other countries in similar situations seems to indicate that universities tend to maximize their revenues as far as physical space allows—every available space in a university is filled with a student.

With public universities granted the right to admit fee-paying students, the Romanian government launched a major battle between public and private universities for the limited number of students able to pay for their studies, a battle in which private universities can only expect to win second place. The interests of public universities are supported by the accreditation agency, which advises Parliament about which universities should exist and which should not. While some present the NCAEA's reporting to Parliament as a step toward decentralization, this is not self-evident. The existence of a university being treated as a legislative issue is probably better explained as a sign of the over-bureaucratization of society than its democratization. What suffers most is the possible diversification of higher education in Romania, something it badly needs in an era of uncontrolled growth, together with a consistent policy that allows a variety of institutions with variety of missions and quality standards to operate without necessarily catching up with largely irrelevant foreign standards that remain out of reach anyway.

Diversifying Romanian Higher Education
The introduction of private higher education certainly brought some diversity into Romanian higher education. It attracted nontraditional students, sometimes to nontraditional programs. And while many now accuse them of copying the programs and structures of public universities, arguably this is a direct effect of the so-called quality assurance policy—imposing public university control on private universities. Private universities are forced to copy those who set the quality standards to survive, and not always to the benefit of the students. Given Levy's distinction,[37] while the Romanian government understands private higher education to be privately funded and governed higher education, it also expects it to fill the educational mission set by the government. Here is the source of potential conflict. There is no need for the government even indirectly to regulate the content of private university studies as long as it is not anti-constitutional and all the parties involved have complete information on the true status of the institution, faculty, program, and the qualification to be received.

It would be unfair to say that the Romanian government is not trying to diversify higher education. It recognized that the quantitative expansion of higher education within the framework laid down under the state-socialist regime, would not meet the expectations that society has for it. Currently, the only distinction runs along disciplinary boundaries. Higher education in Romania should be diversified further. The question, however, concerns the nature of the dimension in which this could be achieved. One of the models proposed offers four dimensions: (i) public–private; (ii) academic levels—first degree, master's, and doctoral studies; (iii) course attendance—full-time, part-time, distance; (iv) territorial reference—national and local.[38] The overall reform is expected to be implemented within the limits of the university. The term *university* is interpreted in an essentialist way where "... the university should have par excellence the vocation of the universal and transdisciplinary relations" and "the term 'university' should be seen as a trademark, answering for a set of demands which must include the status of academic accreditation, multidisciplinarity as a result of the number of disciplines and the possibility of their interaction..." The latter quote, hinting at interdisciplinary studies, is reminiscent of Korka who argued that students' enrolment in two different faculties or universities at the same time would promote interdisciplinary training,[39] one of the curiosities of post state-socialist higher education policy thinking. Without discussing the Romanian government's limitations in developing higher education policy any further, the current thinking about diversification is confined by a rather narrow meaning of the term "university." Smaller institutions will eventually merge with large university conglomerates to bring more uniformity to the system. Policy makers ignore the lessons some Western countries learned in the 1960s and 1970s. Clark makes it very clear that massification of higher education inevitably brings more diversity, "Attempting to cling to old structures and classic traditions, hence finding out the hard way that mass higher education is not simply élite higher education written five or ten times larger, the universities have found themselves in an awkward muddle."[40]

Imposing a rigid definition of university on Romanian higher education ignores the level of diversification required. In one way or another, that leads to the creation of institutions or units either outside the borders of formally recognized higher education or which are forced to ignore existing rules, or both. From recent history, we know that enormous social demand for higher education will be met unless totalitarian rule is reestablished. The best that can be done here is to allow higher education with a wide range of profiles and quality to emerge, whilst answering the necessary information that is available. Political statements about compliance with *international standards* do more harm than good, although they might boost self-esteem temporarily.

One way of conceptualizing the overdiversification of higher education, in conditions somewhat similar to those in Romania, is found in de Moura Castro and Levy's recent work on Latin American higher education. They develop a model that divides higher education into four functional categories: academic leadership, professional development, technological training and development, and general higher education.[41] Romanian higher education, as many other state-socialist higher education systems in Eastern Europe, provided professional and technological training for specialists to fill clearly defined spots in the over-bureaucratized industrial system. Academic leadership was only narrowly covered and only to the extent that the system reproduced itself at the previous level. General higher education did not openly exist; although some areas, teacher training in particular, served this purpose despite official policies. Immediately after the 1989 revolution higher education institutions tried to shift toward academic leadership, while society at large sought to push it toward general higher education. To function successfully Romanian higher education should develop in both directions. The academic leadership should train large numbers of faculty for higher education, as well as other intellectuals. General education should meet the immense need for higher education. It is an error to have the entire system devoted to producing academic leadership.

Evidently, Romanian higher education cannot initiate reform from within. One of the obstacles most frequently mentioned is the attitude of faculty. Although some policy observers in Romania believe this to be basically a psychological issue, solvable by enlightenment among a conservative academy, low faculty morale can largely be explained by the miserable economic conditions. "Their [the students'] teachers, however, are accustomed to fighting barriers, real or imagined, and when they do not exist, they must be invented. The dominant aspiration of professors is 'survival': inside the institution—by the preserving their own teaching load and even acquire some additional ones, and outside the institution—for the sake of a decent living standard, through accepting as many jobs as possible in exchange for an acceptable salary level."[42] While the issue is mentioned time and again in effect, reform works against both change and diversification. Instead of creating more flexible structures, current plans uphold a rigid concept of the university. Beneath it, the internally diverse, massive, and impoverished higher education system will be impelled by a conservative faculty toward a narrowly defined academic mission. It remains to be seen whether official political discourse, the primary purpose of which is to create a positive image for Romanian higher education internationally, will eventually converge with that agenda.

Of crucial importance for Romanian higher education is to develop its sectors of professional training, technological training, and development. Although much of the university faculty seems to despise the recent *polytechnic* training, developing a modern economy and production without it is scarcely possible. This demands structures that relate this type of education to the needs of the labor market. Current practice, in which approving a study program is a matter for national legislation, permits neither sufficient flexibility in these sectors nor the logic of the knowledge economy. When knowledge becomes a commodity, confirming it through national legislative process for the purposes of higher education assures only that higher education permanently lags behind the present-day behavior of knowledge. Technical and technological fields of training should be able to move faster than that. Nor is, however, the issue of general higher education any the less important: "with an under-developed labor market and sluggish pace of institutional, administrative and legislative transition, it is advisable that academic curriculum should promote wide-scope education, capable to open up new opportunities towards future poly-qualifications."[43]

Thus, while need is recognized to a certain extent, the current institutional structure does not support the creation of a general higher education subsector that is able to uphold a wide variety of quality standards and students who would be normally labeled *substandard*. Developing such a sector might be a reasonable alternative to the current policy of letting general higher education flourish under the guise of other purposes. To call it training for academic leadership, however, is particularly counterproductive.

Conclusion

A dozen years after December 1989, Romania remains among the poorest countries in Europe. One may certainly argue over the influence the reform strategies of the governments since 1989 may or may not have had—macroeconomic data reveal only further decline. Each new cabinet entering office expects radical change to occur *next year*. Policies that would lead society into radical discontinuity with the past seem absent. With every year that passes, poverty among the population grows and with it, the task of mobilizing society for change becomes even more difficult. Membership of the EU seems to be the last lever of reform that could possibly bring additional resources to the country, the default of which seems to spell decline. Boia gave a frank account of Romania's political agenda: "Many think of the advantages, especially from a material point of view, but prefer to ignore the structural transformations which such an orientation imposes, the need to rethink political and cultural reference points, and the inevitable limiting of national sovereignty. They continue to hope for a Romania integrated, but at the same time 'untouched' in its perennial values."[44] The ultimate aim of the nation is to become a part of the mythological *abroad* and gain access to its material benefits. To become *foreigners* en masse assuming, for example the role of affluent German tourists, is seen as a solution to many of the current difficulties.

Thus, Romania is caught in a vicious circle: significant changes are impossible without massive additional assistance. No assistance is granted without changes occurring first. While delaying EU membership causes only further suffering to the Romanian people, it is difficult to argue with those who say that without reforms first, external assistance would feed only local bureaucracy and corruption. According to a recent World Bank report, Romania is a society far more corrupt than others in post state-socialist Central Europe.[45] Its companies spend more than 3 percent of their annual revenues on bribing public officials. Romania's level of corruption is comparable to the republics of the former Soviet Union in Central Asia and the Trans-Caucasus.

The conflict between an urgent press for reforms and their apparent impossibility is even more noticeable in higher education. Over the past decade, public higher education has become a massive social safety valve, so much so that the largest share of the national higher education budget is spent on student stipends and various subsidies. Demoralization among both students and faculty is a logical outcome of making higher education function as a massive social service. In this setting it is surprising that governmental agencies and analysts close to them still turn out a discourse expected to demonstrate that Romanian higher education meets *international standards*. Yet, the country spends approximately US$100 per annum on teaching a public university student, that is 2–3 percent of the amount spent by countries with which Romanian higher education compares itself with. Somewhat surprisingly, those who recognize the current difficulties expect the solution to come through a radical change in the *mentality* of faculty and students. Marx would certainly have had something to say on this. However, even without necessarily subscribing to the materialistic camp of radical leftist thought, without a reasonable economy the perspectives for a renewal of mentality remain bleak.

Yet, there are still economic mechanisms that drive Romanian public higher education deeper into crisis. The current university sector, with its rigid if not somewhat arbitrary standards, cannot accommodate 30 percent or more of the age group with all its diverse needs and backgrounds. The private higher education that emerged massively in the early 1990s could have brought necessary variety into Romanian higher education without great cost to the public purse. Here, the government was bereft of any reasonable policy that could have used this initiative to public benefit. Subjecting private universities to bureaucratic control riddled by orthodox and demoralized public university faculty has greatly curtailed its chances of absorbing massive demand for general higher education. Accepting current reality may not be either a pleasant or politically productive step. It might, however, help to draft a realistic plan for more gradual reforms that combine minimal social support for needy students with the involvement of private funding, as well as providing mass higher education over a wide range of subjects and quality standards. Agreed not all would meet the highest international standards. However, the benefit of a more flexible policy is that it would eventually allow inter alia, some real academic excellence to emerge.

Chapter 3
Thirteen Years of Higher Education Reforms in Estonia: Perfect Chaos[1]

The extent to which Mr. Gorbachev's *Perestroika* project was a success or failure is hard to assess. The scheme, which began with the uprooting of vineyards to change the nation's bad drinking habits, led through partial economic reforms and rhetoric on new thinking (all following the maxim "more socialism, more democracy"[2]), to the collapse of a system that even in its very last days was intended to show the whole world the best practices in many matters, including thinking. The final fall came, as Todd had predicted many years earlier,[3] in the failure, or perhaps the ultimate success of the communist system—changes in the respective societies were well on the way. By August 1991, the point at which the Soviet Union formally disintegrated, Estonia had already developed a whole new higher education subsector in private universities that had begun to emerge in 1988.

Social, legislative, and higher education changes stand in complex relation to one another and, for one obvious reason, the weakness of the actors,[4] enjoy little coordination. Attempts to establish more harmony through legislative means have not always been successful because the logic of legal transplants[5] has clashed with the interests of particular groups, for example senior university faculty. This chapter offers an outline of Estonian higher education reform over the period of more than a dozen years that now separates us from the first attempts to break free of the straitjacket of structures and regulations of communist education in the erstwhile world superpower called the Soviet Union. Unfortunately, this essay is one of the very few attempts that have been made to understand the meaning and implications of higher education reform in Estonia, which, unlike other East European countries such as Hungary, still lacks informed public discussion on higher education issues. It would be fair to say that Estonia has not yet been able to provide a critical analysis of the status and future of its post-communist higher education that has, despite far-reaching legal reform, closed itself off from society's intellectual needs and elevated its own status without much external accountability. Among the higher education community, self-promotion is preferred to critical assessment of the past. The "great leap forward" type of utopian program that envisions a radical leap into virtual reality prevails over step-by-step realistic, gradual reform programs. The reality of Estonian higher education could be best described, using the terminology of physics, as that of *cooling-out*—being divided among some 50 institutions it is disappearing as each of its parts flies in a different direction in an ever-expanding universe, a result of a lack of any effective system-wide coordination.

Early Reforms

The Soviet higher education system was fairly uniform in the late 1980s, despite the fact that its 840 higher education institutions were supervised by almost 50 ministries, state committees, and so on.[6,7] Of the total, the Soviet Socialist Republic of Estonia hosted seven higher education institutions, although, because one of them was a military academy unrelated to either local culture or a locally recruited student body, we usually refer to six higher education institutions in Estonia's Soviet period. From today's perspective, these were university-level institutions generally offering five-year degree programs, mostly in fields related to science, engineering, agriculture, and teacher training.[8] While only one of the institutions was officially called a university, their legal status was equal and all were directly controlled by government agencies with no need to deal with issues like planning, budgeting, or curriculum development. These were all dealt with at the levels above the individual institutions. One can estimate that this system accommodated approximately 20 percent of the respective age cohort in Estonia.[9]

Following a now-familiar pattern, one can identify three periods in late and post state-socialist reforms of Estonian higher education. The first period lasted from 1988 until about 1992 and was a period of chaotic, individually and institutionally driven change whilst still within the Soviet Union and its agonizing economic and political realities. During this period, the main divisions of the new and diversified higher education system emerged. Changes during the second period, between 1992 and 1998, took place within an independent nation-state attempting to establish a new legal system and an independent political life, as well as a policy for higher education. During this period, the term *nation-state* gained particular importance when the Estonian government decided to run the multiethnic society with one single language. Consequently, a large non–Estonian-speaking community was denied both access to higher education in their native language and citizenship, which is based on an Estonian language proficiency requirement. Estonian higher education has been monolingual since the early 1990s. Estonia is not a nation-state to the extent to which some would like it to be, but its higher education serves a single ethnic community. Russian-language tracks in Estonian higher education were closed in the early 1990s as the country's independence from the Soviet Union was restored.

During this second period, the higher education sector was consolidated through an active legislative process of governmental decree and the drafting and amendment of the laws that regulated both activities in the educational sector and the operation of educational institutions. Serious efforts were made to bridge the gap between the reality of what had occurred in higher education during the first period and the extant normative basis. It should be noted that, despite much effort, a great deal of what is going on in higher education still bears signs of voluntaristic decision-making by individual high officials and a lack of strategic vision in the development of Estonian higher education.

The third period, which began in 1998 and still continues, can best be characterized as a state of drifting. Neither government nor the higher education community itself has been able do attribute a purpose to Estonian higher education.

The sector is growing chaotically in many directions, most recently toward a growing number of vocational schools that have opened higher education tracks. The Ministry of Education has spent most of its energy in self-reformation and, after 13 years, is unable to diagnose the status of higher education. The guiding logic in this may well be that when Estonia enters the EU in a few years time, junior civil servants from Brussels may come and sort it all out. So, why bother?

Around 1987, criticism of the official educational establishment grew alongside a massive growth in openly expressed dissatisfaction with the Soviet regime and demands for independence from the Soviet Union. The higher education sector was accused of serving the political interests of the totalitarian state and even of constituting a part of its brainwashing system, of being dishonest and hypocritical, as well as having little respect for the principles of academic integrity. Needless to say, military training and Marxism–Leninism occupied the largest share of the higher education curricula[10] and for example, turned technical institutions predominantly into military party schools. In the late 1980s, demands from students for the abolition of military training in higher education institutions drove one of the largest higher education reform initiatives in Estonia.

The reform of Estonian higher education began in 1988—three years before the final collapse of the Soviet Union—as soon as the policy of *perestroika* had created the first rudimentary frameworks within which private economic and civic initiatives could be expressed without the threat of immediate interrogation by the infamous Soviet security agency. Gorbachev's administration recognized the need to promote private initiative in the context of an increasingly miserable national economy. So a new format for this, cooperative enterprise, was legalized in 1986. Although it was not really supposed to extend into the educational field, a group of young people, some of them intellectuals and others less so, established the Estonian Institute of Humanities (EIH) as a cooperative enterprise under the auspices of the Estonian Association of Writers in 1988. This private liberal arts college attracted massive public attention during the first years of its activity, along with students who, for various reasons, had distanced themselves from the communist establishment and its reproductive practices through the official higher education system. It also attracted many visiting faculty from the West whose very presence was one of the first signs of the changes to come.

As the first decade of its activities passed, the EIH lost almost all its competitiveness in Estonian higher education. One might think that this was the result of not very professional management, a lack of strategic vision, and the ability of traditional higher education institutions to capitalize over the long term on their formal status and conventional closeness to the state—whatever it was or is. Nevertheless, the main value of the EIH has been that some two dozen other private higher education institutions have been established following its example. In 1989, the second private higher educational institution, the Estonian Business School, began offering BBA and MBA programs. This project was clearly characterized by a desire to cater to the emerging class of the newly rich and their children, pushing higher education to the borders of entrepreneuralism. This was the first attempt to make a profit from offering higher education in Estonia.

Parallel to the establishment of new, intellectually or socially exclusive higher education institutions, another interesting group also emerged at this time—institutions that were considerably more interested in collecting fees than facilitating students' learning experience, often referred to as *diploma mills*. These were higher education institutions that were officially unrecognized (possibly even unrecognizable because of their lack of teaching capacity) that attracted those student candidates who were unable to find a place in public higher education or meet high fee requirements at elite private institutions, for attractively labeled programs in law or business administration. Although their fee levels were usually modest they often failed to offer anything more than an unrecognized degree certificate. In 1994, the results of a labor-force survey indicated that graduates from at least six formally nonexistent higher education institutions had already entered the labor market.[11] The Ministry of Education had declined to issue a license to operate to some of them while others had failed to even apply for a license. By early 1996, eight private higher education institutions had been licensed while the number of unlicensed ones remains unknown for obvious reasons.

In the context of the changes traditional higher education institutions and the Tartu State University in particular felt, on the one hand challenged by public criticism concerning their collaboration with the communist establishment, if not actively reproducing it, while on the other hand they were also dissatisfied with their status as a part of the centrally planned economic system. This status, whilst having had the benefit of offering lifelong job security, often in exchange for political loyalty, had failed to satisfy those faculty members who had higher than average academic or economic ambitions, the latter in particular, as contacts and comparisons with Scandinavian universities were close at hand. The aim of many in academia can be described in terms of the restoration of the "Ivory Tower" type of research universities following the best nineteenth-century Humboldtian examples. The reform rhetoric in the early days was often mixed with phrases about the revival of the true, medieval European university and a return to the roots. Needless to say, much of this stood in sharp contrast with movements in late twentieth-century European higher education. The ideal of the Estonian university of 1989 can be best described as a strange mixture of medieval idealism and Scandinavian welfare.

In 1989, the then Tartu State University broke the yoke of the old establishment. The Council (senate) of Tartu State University adopted new university bylaws that abolished the word *State* from its title and declared the university to be academically autonomous. In June 1989, the council of the State Committee for Education of the Estonian Soviet Socialist Republic discussed a bill setting out new regulations for the University of Tartu.[12] The council's acceptance of the bylaws, which included an explicit statement on the university's autonomy, set a precedent. Such principles were not only in violation of the totalitarian practices of that time but also of Soviet legislation.

In 1989, rumors had already started to spread that a new university law was under preparation and this forced the other five state higher education institutions to follow the steps taken by the University of Tartu. A particular concern here was the fear that the University of Tartu would be able to capitalize on its privileged status and receive preferential treatment, particularly extra funding, while the growing inflation of the Soviet currency had triggered a rapid deterioration in the economy and created massive public sector funding difficulties. Shortly afterward, all five state higher education institutions changed their titles and became universities. The Polytechnic Institute became the Technical University, the Pedagogical Institute became the Pedagogical University, and so on. For some of them this was the only step toward reform for several years to come. The State Assembly (the Estonian Parliament), however, only adopted the University Law in December 1994.

Changes made within the educational system during those years were often of a rather random nature. An interesting reform precedent in public education was set in the secondary education sector. In 1989, the Special Boarding School for Sports in Tallinn (a secondary school for highly talented young athletes) was renamed as a *college*,[13] a step that revealed a desire for higher status and closer links with nonuniversity higher educational establishments. The fact that only a year later the school was again reorganized, this time to become a *gymnasium*[14] demonstrates the desire to break off from the Soviet system without really knowing how to do so. Such initiatives were largely inspired by dissatisfaction with the existing situation, rather than by a clear strategy for the future. The use of words such as *gymnasium* and *college*, which had basically no local cultural reference, in naming new institutions had no inner meaning; they simply referred to the desire for something different, new, or more legitimate than the educational systems set up by what the official history now terms the *Soviet invaders*. In the years that followed, the word *international* in institutional titles became one of the most important signals of the discontinuity with the communist past.

During the same period, another massive force began to challenge the higher education establishment from below, eventually creating a binary divide in Estonian higher education. It was the drift of the former vocational schools, under pressure from economic and political insecurity, toward higher status that introduced a new sector in Estonian higher education—vocational higher education. Using the example of the structure of German higher education, particularly the *Fachhochschule* [trade or technical college] as the target and capitalizing on sweet memories of Estonian historical connections with Germany, vocational schools began to boost their status. Just as the University of Tartu had taken the initiative in amending its bylaws, so the Tallinn Technicum of Construction and Mechanics decided to become a part of something that did not formally exist—the vocational higher education subsector of Estonian higher education—becoming the Tallinn Higher Technical School. As several other schools followed the initiative, a new subsector soon emerged. As it had been with private higher education, this was clearly outside of the existing legal framework. However, as the entire country was under reconstruction civil servants in the Ministry of Education had hardly any idea of how to deal with such initiatives and certainly had no time to get involved. Seven such institutions rapidly emerged. Subsequently the Ministry of Education issued a decree to temporarily regulate the activities of institutions that had decided to take their fate into their own hands.[15] Needless to say, faculty members in these institutions became *professors* instead of *teachers* as their former status had been. This also counts as a drive toward change.

To sum up the main events during the early period of reform, one can identify three major events, each of them marking the beginning of a subsector in the new, diversified higher education system. Upon them lies the structure of the new Estonian higher education system, as well as the structure of higher education legislation. These three events were: the establishment of the Estonian Institute of Humanities—the first private higher education institution in 1988; the adoption of new university bylaws by the council of the Tartu State University and their subsequent approval by the State Committee for Education in 1989; and the transformation of the Tallinn Technicum of Construction and Mechanics, a typical Soviet-era vocational school, mixing secondary- and postsecondary-level programs into a vocational higher education institution in 1992.

The Consolidation of Estonian Higher Education

The reforms of the period 1992–98 were driven by two main concerns. First, the earlier changes that the system had experienced were unsystematic and guided by an utterly disrespectful attitude toward law and order, Soviet or other, from educational institutions, as well as individuals who wished to become the rectors of their own universities. Higher education required a new legislative basis that would allow diversity whilst at the same time introducing a certain level of regularity, as well as putting an end to the extreme voluntarism that the system had experienced earlier. Second, the government needed to restore a certain level of control over the operation of higher education institutions. One sign of the lack of any perceived alternative was that some of the most senior civil servants envisioned restoring central administrative control similar to that previously exercised in the Soviet Union. Others preferred the design of a governing structure where significant power would be delegated to so-called buffer organizations. The result seems to be a compromise where buffer organizations have not been created and the Ministry of Education is not able to perform all its functions either.

The solutions to these problems were sought from two clearly different sources. The fundamentals of legislative reform were derived from the continental European legal tradition of the civil code as the most important law. New administrative practices, however, were guided to a significant extent by the state-socialist practices from which many civil servants had acquired their skills and values. It is therefore unsurprising that the results have been mixed and often contradictory. Discussing legislative reform in Estonian higher education one should also bear in mind that the normative acts adopted in 1989–94 were often driven by a legal thinking that differed from that which motivated those adopted after 1995. The purpose of the former was to provide a legal basis for diverse groups of institutions, latter legislation had its roots in the new civil code adopted in December 1994.

The continental European juridical tradition that was once the source of the Baltic Special Law that the Russian Empire adopted in 1884, once again became the cornerstone of the Estonian legal system. It rapidly gained an importance resembling that described by Watson, "In European countries, the civil code is always regarded as the basic law—it is on this that most attention is focused—and the other codes are to some extent treated as secondary."[16] For the Estonian population, enactment of the new civil code adopting *The Law of the General Part of the Civil Code*[17] by the State Assembly was also the ultimate sign of the discontinuation of the Soviet legal system. The enactment of the new civil code in 1994 necessarily led to the revision of the laws regulating the educational sphere that were adopted in 1991–94. For example, the *Law of the Private School*[18] initially adopted in 1993 had been completely revised by 1998.[19]

The legal nature of higher education institutions in Estonia is defined by *The Law of the General Part of the Civil Code*. This legislation radically redefined the previous State-dominated legal landscape and provided a new classification of subjects of the civil code in Estonia defined as *persons*. Section 5 states that subjects of the civil code, the persons, are divided into *physical* and *legal* persons: physical persons being individuals and legal persons created *according to the requirements of the Law* (§5 Art. 3). Legal persons are divided into *Legal Persons of the Public Law which are created in accordance with the law in the public interests* (§6 Art. 2) and *Legal Persons of the Private Law created in the private interests* (§6 Art. 1).

From the perspective of the transition between the old and the new civil codes, it is important to notice that the latter states that *the State and the Local Governments participate in the legal relations as Legal Persons of the Public Law* (§6 Art. 3). Transition can be interpreted as dividing the former large, actually totalitarian state into various legal persons, either under public or private law, and having the new, circumscribed state as an equal subject with other *persons*—one legal person of public law among many others. The state universities were separated from the state and became legal persons of public law along the same lines.

In summer 1998 the enactment of *The Law of Vocational Higher Education Institutions*[20] regulating the activities of the ten state-run vocational higher education institutions was the last step in establishing a new legal environment for higher education institutions in Estonia. *The Law on Private Schools* that regulates the activities of private educational institutions at all levels, including the four private universities and 14 private vocational higher education institutions, was enacted in 1993 and a completely revised version was adopted in 1998; *The University Law*[21] regulating the activities of six public universities came into force in 1995; and *The Law of the University of Tartu*[22] that addressed the uniqueness of the only classical university in the country was adopted in 1995. Issues closely related to research and development activities are regulated by *The Law on the Organising of Research and Development Activities*,[23] the current version of which was adopted in 1997. The need for the latter can be explained by the existence of a large research system outside of higher education, the Academy of Science Research Institutes. After some attempts at merging with these, the universities backed out because of the high cost of running these inefficient, unproductive, and bureaucratic organizations, and now the entire sector may gradually be disappearing.

Within the new higher education legislation each institutional type has its own law. The enactment of *The Law of the University of Tartu* in 1995 and its subsequent revisions, which have significantly decreased the scope of the law, may, however, be interpreted as unsuccessful attempts to create a privileged institutional category, the *national university*, comprising a single university. This once more addresses the deep conviction of the University of Tartu that it is more than just a university among many. It also indicates the level of political influence a single university can exert in a small country with regard to the national legislative process. In 1994–95 the

University of Tartu was able to persuade the Ministry of Education to draft a law to protect its special status and to bring it to the State Assembly for a vote. From a purely legal point of view there did not exist a need for such a law.

Most interestingly from the legal point of view is the comparison between the university and vocational higher education sectors and their respective legislation. This allows an analysis of the extent to which legislators have been able to make sense of the principle introduced in the new civil code that distinguishes between different types of legal persons. Following this logic the state-controlled Vocational Higher Education Institutions (VHEI) still constitute a part of the state apparatus, while the former state universities enjoy the status of legal persons of the public law.

The Implications of the New Legislation
With the logic of the civil code, which distinguishes between the state, public, and private entities, being translated into the higher education legislation, three groups of institutions are clearly separated. Public universities as legal persons under public law enjoying significant administrative autonomy; vocational higher education institutions constitute elements of another legal person, the state, having very limited autonomy while private higher education institutions are considered as private entities.

Defining VHEI as state higher education institutions means subjecting them to the direct control of the Ministry of Education. The intention of the Ministry of Education, as well as the logic of the law, is to control them in the way the previous political regime controlled its whole higher education sector. Institutions of this category are established by government decree. The Minister of Education contracts the rector, as well as approves the institutions' annual budgets. Furthermore, the Ministry of Education controls the academic activities of a VHEI—the opening or closing of study programs and the determination of admission quota of students for every field of study. Quality control comes under the Higher Education Evaluation Council that deals with all institutional types.

The university law regulating the public university sector clearly indicates that a certain level of *de-étatization* has been implemented. Through a series of amendments between 1995 and 1999 the logic of the civil code has been translated into the law, particularly as related to governance and administration of public universities. While, for example the initial version of the law regulated the nomination and contracting of rectors in a similar way to the VHEI, later amendments have made it an internal university issue. The University Council elects the rector and the oldest member of the council signs the contract on the university's behalf. While the law does set minimum requirements for the faculty positions, appointment of faculty and staff positions is also an internal university issue. A major difference from the Soviet system is that the universities also award higher degrees, including doctoral degrees.

From the formal legal point of view, this is exceptional autonomy in the continental European context. The whole university faculty and staff are contracted to the university, unlike many continental European countries where the faculty are civil servants and, at least formally, nomination of full professorship is the prerogative of the head of state, for example as it is in Finland.

The total withdrawal of the state from the governance of public universities also has a negative side. While universities are funded from the public budget, no mechanism of public accountability is in place. Universities do not have trustees with a mandate reaching beyond vaguely defined advice to the rectors of universities. It may well be, therefore, that public universities have at least partially achieved the long-desired Ivory Tower status. An inability to build effective mechanisms of public accountability into the governing structures of legal persons of public law seems to be one of the major problems Estonia is currently facing and concerns not only universities but also other entities with the same legal status, for example the public television network.

It cannot be denied that there may be universities of approved excellence that can work at their best without external intervention. This may not, however, be the case in post state-socialist Estonia where mass higher education institutions have not been able to radically improve either their teaching or research capacity over the past decade. In this circumstance the new legal status serves primarily as a shield against establishing public accountability. While the University of Tartu has been championing complete independence for universities, beyond receiving lump-sum allocations from the public budget, having such principles translated into national legislation also has its negative side. Some of the institutions (particularly the Pedagogical University of Tallinn, formerly the Pedagogical Institute that represented one of the lowest classes in Soviet higher education) do not have the capacity to organize a consistent educational process alone and in the current conditions there is no structure in place to help them to do so. The quality assurance mechanism has become a process of political negotiation over formal accreditation and fails to help higher education institutions to improve the quality of the education they deliver.

Financing of Higher Education
One of the most difficult problems facing the Estonian government is financing the large and inefficient higher education system it inherited from the Soviet era. It seems quite obvious that the post-communist economy is unable to finance universities, some of which have faculty/student ratios as low as $1:4$.[24] At the same time, attracting funding from other sources is also difficult as the law stipulates that higher education should be offered free of charge. Over the past decade the government has made several attempts to introduce direct tuition fees, an initiative that usually causes public unrest despite the existence of a student loan system in the country.

The government's solution has been to distance itself from funding by introducing a formula funding mechanism through which public monies are distributed to universities according to student places, weighted by fields of study (humanities, engineering, medicine, etc.) and level (bachelor, master, etc.). The mechanism has at least two shortcomings. It is obviously unable to generate the additional funding public higher education so badly needs. Furthermore, the various weights the funding formula applies are more a reflection of the power positions held by particular universities at the time when the universities and the Ministry of Education negotiated them than any objective *scientifically* established criteria to which it pretends to be related.[25] To generate additional funding in the mid-1990s universities began to admit fee-paying students on the top of the state admission quota, calling the activity an *additional educational service*. The Ministry of Education, unable to legalize tuition fees and avoiding direct confrontation with universities, temporarily tolerated this as it tried to find a legal compromise that would allow universities to charge fees for a certain population of students whilst at the same time maintaining the official, free-of-charge higher education policy.

The compromise was to rename the student admission quota as the *state order*, allowing universities to admit an additional 20 percent of students on a fee-paying basis. The actual number of fee-paying students, however, seems to be significantly higher than that. This compromise, which among other things has triggered some charges of corruption as a criteria for becoming a state-sponsored student, are not always explicit or respected and is reminiscent of a particular policy initiative in the Soviet Union some 20 years earlier. When the Soviet Union faced a severe food shortage Mr. Brezhnev and his *Politburo* encouraged the collective farms to produce more by allowing them, after having met the requirements of the state plan, to sell any surplus in the marketplace. Introducing similar mechanisms into a largely market-driven society almost inevitably leads to corruption. However, society at large seems to be quite a distance from understanding that state-funded higher education is not free of charge anyway or that certain degrees lead to significant private benefits, so tuition fees may not be quite the evil that many perceive them to be. They would also allow the transformation of a significant part of public funding into scholarships to be distributed to particular students according to explicit criteria, means, or merit-tested, making the entire financial sphere in higher education significantly more transparent.

The Academic Content of Studies
While higher education institutions, particularly universities and private higher education institutions, have significant autonomy concerning administrative and governance matters, the Ministry of Education seems to believe that it is in the best position to decide upon the academic content of studies. Every new teaching program at universities or the VHEI have to be formally approved by the Ministry of Education, which also maintains records on all study programs, as well as all individuals to whom higher education institutions, including accredited private institutions, have conferred degrees.

The most significant element of state control over the content of higher education studies is the quality assurance mechanism launched in 1995. Quality assurance, which in the Estonian context means compulsory accreditation of all programs in all types of higher education institutions, as well as all institutions themselves, is a strange mixture of the new quality movement in European higher education and Soviet bureaucratic practices. The process is run by the Higher Education Evaluation Council (HEEC), a unit under the auspices of the Ministry of Education that once was meant to be a buffer organization, but came to be just another office within the Ministry. The HEEC has 12 members primarily representing the higher education and research community. The members are proposed by the Minister of Education and approved by the government.[26] Its decisions are offered as advice to the Minister of Education who either accredits an institution or a program or does not. A negative decision is an immensely discrediting step leading to severe consequences: the closure of the program or institution as a whole by administrative means.[27] Formally, all higher education institutions in Estonia should pass accreditation. In reality the case seems to be that the HEEC seems to represent the quality perceptions of traditional universities and may be biased against private institutions. Given that the mission of the latter group is often alternative, evaluating them against the same criteria as traditional universities may not be entirely appropriate.

Lack of public policy in general does not make a theory about public universities conspiring with the Ministry of Education against private education look feasible. However, one has to admit that the way the Ministry and the public university faculty run the quality assurance process makes private education its primary, if not its only target.[28] Representatives of the Ministry of Education, along with public universities, have recently started expressing concerns about there being too many higher education institutions. If steps were taken to close them this would inevitably drive the fee-paying students to public universities, which makes one consider whether the primary concern of public universities is in securing their own funding base or the broader educational needs of the country. One can also pose a more fundamental question about who really decides on the academic matters delegated to the Ministry of Education. The answer is either no one, or it is the most politically powerful universities, their rectors, and senior faculty.

The main document on the basis of which a decision on accreditation is made is a self-study report prepared either by a study program or an institution as a whole. Establishing the process in this way allows quality assurance in Estonian higher education to be presented as a part of the Westernization program. However, while, for example a similar process was introduced in the Netherlands in the 1980s as a primarily formative, improvement-oriented measure, in Estonia it has taken a clearly summative form providing a pretext for the most threatening administrative decisions.[29] This has turned the whole process into a highly political negotiation between various higher education institutions and academic groups.

Private Higher Education

As in other East European countries, there were two principal reasons for the emergence of private higher education in Estonia. First, state-controlled higher education was unable to offer the education that the rapidly changing society needed; and second, there were a number of people, quite often coming from traditional universities, who had the knowledge and initiative to offer the courses that the market demanded. Looking more closely at the structure of the market one can discern three types of demand: the demand for alternative liberal education—a demand first met by the Estonian Institute of Humanities established in 1988; the demand for studies in fields that state universities did not offer, for example business administration— first offered by the Estonian Higher School of Commerce established in 1989; and the demand for exclusive elitist environments, offered by the Concordia International University established in 1993. There is also a hidden demand for diplomas or certificates of any kind for minimal effort and outlay, a demand usually met by institutions that pretend to be providing exclusive, high-quality education.

But it is also the case that all higher education institutions in Estonia, even the ones that were the most established, greatly exaggerate the level of their academic excellence in the public relations campaigns they present.

Estonia has one of the highest ratios of inhabitants per higher education institution in Europe, or perhaps anywhere. More than 40 higher education institutions per 1.5 million inhabitants is certainly more than in Romania, which has the most widely discussed higher education explosion in the post state-socialist Eastern Europe. In 1987, Estonia had six higher education institutions with an annual intake of approximately 5,000 students.[30] By 1996, the number of higher education institutions had grown to 21,[31] while because of the abolition of part-time studies in public universities, the number of students had not yet grown significantly. In 2000, the total number of higher education institutions reached 41, and the total student intake was 12,000.[32] In 2001 there were 48 institutions offering higher education.[33] Out of a total of approximately 56,000 higher education students in the year 2001, approximately 12,000 study in private institutions. However, figures like these are not always highly reliable because of the existence of a certain number of institutions operating without having applied for a license or accreditation. The Ministry of Education also seems to have difficulty in determining the number of institutions offering higher education, particularly as many vocational schools, perhaps an unknown number, have recently opened higher education tracks.

Despite the large number of private higher education institutions in Estonia (at least 19) less than a quarter of the student population are accommodated in them. The student bodies in private higher education institutions rarely exceed 1,000, and their organizations are not always very strong, particularly as they often rely on part-time teaching staff coming from established public institutions. Due to the reliance to the greatest extent, often totally, on funds generated through tuition fees set by the modest living standards in the country, infrastructural development in private institutions is also a serious issue. All of which provides public universities with a good reason to criticize private institutions for their lack of infrastructure, full-time faculty, and so on, and also to write particular, generally unrealistic, requirements into the quality standards. At the same time one has to admit that the learning experience that the low-end private institutions offer is very limited. This does not, however, mean that public education does not offer comparable examples, but it does have the clear benefit of attracting the best students and awarding fully recognized degrees. Many private institutions are caught in a vicious circle that they cannot break out from: limited funding, problems with recognition, bad students, more financial problems, and so on.

Under pressure from the traditional higher education sector and the accreditation mechanism that largely serves its cause, only a few private higher education institutions have been able to maintain the alternative or elitist profile they initially claimed. The best secondary education graduates, particularly from affluent urban families, are competing for a limited number of state-subsidized student places in public universities, and less affluent and less well-prepared students occupy fee-paying positions, including those in private universities. In this respect Estonian higher education follows the pattern of Brazil's.[34]

There is no doubt that Estonian higher education has become quite diverse over the past decade. However, this does not mean that public higher education is of superior quality and that private is of low quality. The case seems rather to be this—as many public institutions have been able to effect little change, some private institutions are doing important work in offering alternative courses and serving students who would not otherwise have any access to further education. Instead of promoting public discussion on higher education and making available information about the various institutions, courses, and so on, the current approach seems to be to let the conservative academia impose its own standards on higher education, and fight alternatives through administrative means.

The Policy Vacuum and Drifting in Estonian Higher Education

In some respects Estonian higher education is in a better position than most of its East European counterparts. This particularly concerns the legal foundations that were established with the comprehensive reform of the Estonian legal system completed by the mid-1990s. The way the civil code defines the institutional landscape has created a favorable environment in which many of the problems other countries struggle with, for example governance and funding, could be solved with relative ease. This has allowed a division to be created that distances public universities from the state without forcing them into the private sector. Many of the former Soviet Union countries that have not undertaken such reforms face a spectacular dilemma in either controlling higher education directly as a part of the state apparatus, which usually means acting in the way the Soviet Union once used to control its higher education, or by privatizing the higher education sector entirely as, for example Kazakhstan seems to be doing without a middle way between the two. Although *privatization* is certainly a buzzword in the post-communist world, the state completely renouncing its role in regard to higher education may have rather severe long-term implications for the intellectual life and culture of the countries that choose such an approach. Needless to say, under such circumstances culture does not always matter.

Estonia, as mentioned earlier, is in a better position. It has created a structure that allows higher education to operate in the public sphere with the state playing its limited role in it. It is therefore even more ironic that despite the legal foundations, Estonian higher education seems to be gaining no benefit from this, and seems to be drifting toward chaos without much policy guidance. While the legislative reform in higher education is an outcome of the comprehensive legal reform that the country undertook after breaking away from the Soviet Union, like many other East European countries Estonia has not been able to pursue active educational policies since the fall of state-socialism. The legal framework, highly systematic and following the good old East Prussian tradition, is probably too regular to ever work properly. The most difficult issue seems to be the lack of a comprehensive interpretation of the civil code for the purposes of higher education. The result is that while certain issues like governance have fallen completely out of public control some others, for example formal control over the content of studies, have been seized by the ministerial officials in a way that exemplifies the continuity between the old Soviet and the new Estonian bureaucracy. However, as the state bureaucracy is relatively small compared with that of the Soviet Union, the many functions brought under the roof of the Ministry of Education are not exercised beyond the most superficial level of filing the documentation. The Ministry of Education clearly lacks the capacity to have a meaningful role in reviewing the content of higher education programs, but it still retains the function of maintaining files of all the program descriptions. In this way, institutional lobbying for individual favorable decisions drives daily decision-making. This has created a picture in which higher education looks far less perfect than the logic of its legal foundations.

The main political agenda of the incoming governments, both after the 1995 and the 1999 general elections, has been about reforming the Ministry of Education itself and not the sector it oversees. In 1995, the main political issue was to divide the then Ministry of Culture and Education into two ministries, one for culture and one for education. The primary reason for this was a leading member of a main coalition party who was interested in occupying the position of Minister of Culture when there was at that time no such Ministry in the country. Separating the necessary structures from the Ministries of Culture and Education took significant amounts of energy and impaired their functioning for a while. After the 1999 general elections, the new government coalition had an even more radical agenda, moving the Ministry of Education from the capital Tallinn to the second major city Tartu some 200 km away into the physical vicinity of its dominant political driving force and client, the University of Tartu. Preparing for the move took more than two years, and in summer 2001 the new Ministry of Education is expected to start operating. The difficulties in running daily activities are obvious when ministerial officials are being laid-off or new ones are hired, while offices are on the move or new structures are being developed. As far as higher education is concerned, the actual picture that has emerged by 2001 suggests that if the next government chooses a similar reform policy, that is, to reform only itself, then difficulties in running higher education for public benefit will escalate even further and the whole sector will drift toward the private sphere, becoming increasingly entrepreneurial and serving the private interests of those who enjoy access to decision-making. But it may also be the case that after so many reforms the structure has lost its capacity to deal with anything outside itself and therefore a new reform of the Ministry after the 2003 general elections would be the only way forward.

In June 2001, the Estonian government approved by decree a document that outlines higher education reform for 2001–02.[35] The fact that such a document has been drafted is itself curious as, according to some high officials in the Ministry, the reform was already successfully completed a few years ago.[36] Therefore, the primary purpose of this document is, as is often the case, not so much to address the problems higher education is facing, as to pacify the higher education community by making promises for more funding *next year*. Although the descriptive and analytical parts of the document are rather limited—two pages of the ten-page document—it does offer a rough picture of the most recent proliferation in Estonian higher education.

In 2001, Estonia has 48 institutions that offer higher education, accommodating a total of approximately 56,000 students. An important distinction that needs to be made here is between higher education institutions and institutions offering higher education. Fifteen of the higher education institutions have university status and 20 of them are vocational higher education institutions. Both groups include public, as well as private institutions following the structure created by the various laws described earlier in this chapter. The most interesting part is, however, that the 13 institutions are not higher education institutions, but they do offer higher education programs. These are vocational schools that have been authorized by the Ministry of Education to offer higher education programs. This is an indication that once again institutions have begun to drift toward higher status and challenge the existing legal framework, as they did in 1988, in an attempt to increase their funding by opening higher education–level study tracks. It is even more surprising that while the government seems to be worried about this development and the confusion over the structure of the higher education sector, each of these institutions offering higher education programs has been authorized by the Ministry of Education to do so. More than that, the outline of every single program is being approved by the Ministry of Education and filed with it according to the existing law. It appears as if the government is not fully aware of what it is doing as it opens higher education programs in vocational schools and then complains that the situation is out of hand, and that the higher education scene is becoming confused and the quality of teaching is deteriorating. It is odd that the government is worried that it is not in a position to introduce a policy that would guide the officials who authorize the opening of new institutions and programs. This particularly concerns the state-owned vocational schools that do not have the capacity to provide higher education. By the end of the Soviet period, higher education institutions in Estonia offered approximately 300 different programs, which was considered excessive. This number had grown to more than 900 by 2001 and exceeded 1,200 in 2003.

The number of higher education students has grown almost threefold since 1989. This figure may, however, be misleading as it includes many of those who were previously counted as secondary school students studying at vocational schools. Evidently, the higher education sector has experienced major growth, particularly since 1996, although nobody knows exactly how much. It is hardly possible that any new student places need to be created. Even if the number of student positions in higher education remains constant, by 2007–08 Estonian higher education will accommodate the entire 19-year-old cohort because of the sharp decline of the birthrate after 1989. Already by 2000, according to some commentators, the number of new entrants exceeded significantly that of secondary school graduates. Even if one considers that many older students have returned to higher education, and that can also cover all students who study at vocational schools that offer higher education programs, this figure is still shocking. There seem to be many reasons why the reported numbers of higher education students are exaggerated: for political reasons such as catching up with OECD standards, as well as economic reasons such as receiving additional public funding.

Although Estonia seems to have problems in developing higher education policy and also in gathering basic information on its performance, meeting Western and international standards still constitutes the very core of its policy rhetoric. Here, the system is giving conflicting signals on its status. On the one hand, the Ministry of Education declares that meeting international standards constitutes the most important goal for Estonian higher education and in particular, the notorious Bologna Process and its targets guide formal higher education policy. On the other hand, many academics declare that Estonian higher education is the best the world has had on offer for some 200 years. Arguments have been made to the effect that not Oxford and Cambridge, but Tartu has been the best European university since the nineteenth century.[37] It is not too difficult to see the aim of such statements—shielding universities against the winds of change.

Estonia is at least as eager to join the EU as Romania, or even more so, and demonstrating its readiness to do so constitutes the core of government policy. The main form this takes is the production of reports that are sometimes only indirectly related to reality. It is an increasingly popular trend among officials to identify *European standards* and then declare that Estonia meets them all. This is also extended to areas in which the EU, according to its founding charter, cannot have a common policy, such as education. Many East European post state-socialist governments, including that of Estonia seem to be attributing a stronger rôle to the supranational European institutions than they actually have. Although the EU, following its principle of subsidiarity, cannot have a common educational policy, Estonia still expresses its readiness to comply with such a policy. This is partly a result of the lack of information. Another part, however, is wishful thinking. It would certainly be a great relief if the EU had a common educational policy and it would fund it. Therefore, there is great positive expectation in trying to find it.[38]

What is taken to be the common European higher education policy is the so-called Bologna Process, which aims at creating greater compatibility between the higher education systems in Europe.[39] The difficulties such a task eventually faces are evident. First, at least formally, there is no agent to carry such a policy across Europe. Second, traditions and status attached to particular types of institutions vary widely from country to country and are not easily overcome by decrees or agreements. One could, for example think of what it would take to establish formal equivalence between the German *Fachhochschule* and French *grande école*. The Bologna Declaration is actually aiming at considerably less—introducing an undergraduate (3–4 years) degree by the year 2010. Still, some of the leading higher education experts, for example from Germany, find it a rather challenging task. The Estonian perspective on the issue is somewhat more positive. It expects to introduce the necessary changes by 2003. Here, one has two alternative ways of viewing such a statement: a higher education system that can be restructured in three years either does not have much of a tradition or many established structures, or one sees a policy replaced by political opportunism. Estonian higher education seems to have entered a phase where the government is able to implement reforms almost immediately without much institutional resistance. The Ministry of Education is busy with permanently reforming itself and whatever policy document it might prepare for higher education, it lacks the capacity to follow up its implementation anyway. Operating in such a manner, there is little connection between higher education and the Ministry of Education although the Ministry formally controls certain key areas of institutional life, for example the content of studies. In reality, at best, the Ministry produces documents for political consumption while the institutions basically do as they please, being accountable neither to the government nor to the public for whose benefit, at least formally, they act.

Conclusion

There is no doubt that Estonian higher education has experienced significant changes over the past dozen years. Even more than that, the very concept of higher education has been radically redefined from postsecondary education to university. Along with this the institutional landscape has diversified and the number of institutions and students has also grown significantly.

The legal framework for higher education has altered as well. Chaotic changes in the structure of the higher education system provoked governmental legislative initiatives in the early 1990s. A second wave of legislative reforms followed from the mid-1990s aimed at bringing the earlier normative acts into conformity with the new civil code. Its logic, however, has not always reflected the needs and problems of higher education. While many contextual issues like governance and funding have been dealt with and virtually all institutions have been renamed, the actual learning experience of students may be not be that different from what it was 15 or 20 years ago. Academia has protected itself against the changing needs of society and laid stress on, for example its historically high international standing. The faculty largely remains the same and operates with limited resources within its own intellectual limitations.

Estonia has been successful in creating a new legal basis on which public higher education could possibly operate. This has not, however, been translated into policies that guide the higher education sector. The fact that the Ministry of Education has taken control of a function that it does not have the capacity to exercise has not allowed the establishment of proper links between higher education and society. The result is that while the government still maintains the illusion that it is in charge, higher education institutions do as they please, often representing the level of the lowest common denominator—maintaining the status quo and opposing an outside world that may eventually ask for a somewhat higher level of relevance. The recently manufactured glories of the nineteenth century may not be that helpful in the twenty-first century.

Chapter 4

Russian Higher Education After Communism: The Candy Man's Gone[1]

The immense and inefficient educational systems inherited from the Soviet Union and its former satellites in Eastern Europe have placed the successor countries in a rather difficult situation. After the Second World War in particular, the Soviet Union developed an educational sector that reflected neither the level of its economic development nor the manpower needs of the country. Having set a single goal, gaining military superiority over the Western World, Stalin's regime directed enormous resources into higher education and research. People such as Voznessenski have considered the ability to redistribute resources on a massive scale by a highly centralized totalitarian regime as a particular benefit of the Soviet system: "the use of the benefits of the Soviet system would allow Russia to gain superiority over the capitalist countries in all development tracks including that of technological development."[2]

The annual number of higher education graduates doubled from the prewar period shortly after the war. During the 1950s and 1960s the Soviet Union increased its spending on research twelvefold. The administrative tools used by the former totalitarian regimes to redistribute limited public wealth could certainly not have continued after the fall of state-socialism and this created massive difficulties for the higher education and research sector as a whole. The collapse of the centrally planned economic system and its industries and the ensuing extensive unemployment and social insecurity have, however, further increased the social demand for higher education beyond the level that had existed until the late 1980s. At the same time, higher education institutions, which have only received sufficient public funds for modest salaries and student stipends over the past decade, are running out of human and material resources. Russian Federation's current per capita spending on a higher education student at the level of US$360 per annum remains well below the level that would allow the system to survive, not to mention its reform, relevance, and quality.

There are certainly strong reasons not to address issues related to the integrity of the Soviet higher education and its real achievements, some of which might already have become clear in the earlier chapters of this book. To put it very simply—because continuity between the old and new higher education is strong, particularly considering the personnel involved, as well as the content of studies—criticizing the past actually involves criticizing the present. Saying that social scientists under state-socialism served the goals of evil regimes is tantamount to saying that social scientists who currently teach in post state-socialist universities lack moral integrity. To expect such self-criticism to spring from within universities is quite futile.

One could possibly argue that Soviet higher education failed on two important counts. First, on moral grounds, it was unable to promote anything that could come even close to free intellectual enquiry. Soviet higher education failed to carry out its universal intellectual mission. The function of the Soviet intellectuals was not to exercise free intellectual enquiry but to confirm and spread the truth as prescribed by the communist establishment. It would not be too much to argue that post-socialist corruption, including that of higher education, has its roots in the immorality of the communist system. This, however, gives no reason to argue that capitalism is by default any more moral than communism, although in general its record looks somewhat better. Second, the size of the higher education system that the Soviet Union once developed stands in stark contrast to the level of its industrial and economic development, indicating a complete failure in the main task bestowed on it by the communist establishment, that is, assuring the Soviet system of industrial, economic, and military superiority over the capitalist Western world. Those who argue that investment in higher education has a direct impact on economic development should give some thought to this, which is not, as the Soviet example shows, necessarily the case.

While the general problems of higher education systems in East European countries are similar (low salaries, lack of investment, faculty and students engaging in multiple employment, administrative rigidity, a growing gap between the formal over regulation and centralization, and the atomization and voluntarism in daily practice) the systems are drifting apart with accelerating speed. The common denominator among the East European higher education systems may soon be no more than that of its fundamental driving force, "minimal utopia-survival,"[3] for there is apparently no power similar to Stalin's or Brezhnev's Moscow, which once had the capacity to control the ideological soundness of not only its own universities, but also those of Prague, Budapest, and a large part of Berlin. Perhaps, one day the United States of Europe that is currently under construction will be able to repeat Soviet achievements and introduce highly centralized bureaucratic controls, and whilst many East Europeans would gladly see Moscow's power simply transferred to Brussels, there is still some way to go.

The position of the Russian Federation and its higher education on the landscape of post state-socialist higher education reforms is radically different from its former COMECON satellites and the three Baltic countries. The latter two groups are making serious efforts, albeit largely rhetorical of late, to catch up with the economically developed countries. Such an approach is politically unacceptable in Russia where officials and senior academics continue to declare that Russian education sets world standards.[4] Generating the myth of Soviet and Russian intellectual superiority, its scientific advances and military might has for decades helped autocratic regimes to mobilize the country's population into nothing less than slave labor under extreme conditions. One has only to recall how thousands of men were sent into Chernobyl in 1986 to fight the worst-ever nuclear disaster without preparation or the means to do so beyond the encouraging words of the communist leadership that a Soviet man was stronger than nuclear radiation. Such a record makes it difficult to critically review the content and processes in Russian higher education or its quality and true intellectual contribution. It is particularly surprising that Russian higher education in its entirety seems to be completely uninformed of the criticisms made of it by Western scholars, and more recently by those in the former state-socialist countries. Although some of it has been translated into Russian, for example Graham's well-balanced but critical book[5] or the recent OECD report, and is thus available to Russian policy makers, for one reason or another it is not a subject of discussion in either policy making or academic circles.

Unlike its former East European allies and perhaps the Baltic countries who in the late 1980s at least briefly faced the issue of their higher education having violated certain central academic values, this issue has not been addressed in Russia. Consequently, there is no perceived need for any reform beyond letting the world know about the true excellence of Russian education. As discussed in earlier chapters, there are similar trends in both Romania and Estonia, although it is more difficult to maintain them in smaller and more open countries with such differences in the scale of economies. Poverty blocks Russia's access to many external information sources and a massive language barrier still separates Russian academics from the rest of the world. A second important characteristic is that other countries belonging to the Commonwealth of Independent States (CIS) have only limited access to Western intellectual and economic resources and still closely follow Russian policies. Accordingly, there are perhaps some 11 newly independent countries that closely follow changes in Russian higher education policy and legislation and ensure that their national systems remain compatible with these, if for no other purpose than to maintain labor mobility. Whatever the difficulties in the Russian economy, with its abundant natural resources it is still doing better than the rest of the CIS. Whilst revising this chapter I read an article in *Newsweek* magazine about Moldavians selling their kidneys for $3,000 to Israeli citizens in Turkish hospitals.[6] Now one may call this "Twenty-first-Century Europe," but it certainly represents something other than that usually meant by such a vocabulary. The poverty in the post state-socialist, former Soviet Union is beyond anything in European understanding, and while Russia is poor the rest of the CIS is much worse off.

The former Soviet higher education system now comprises the core of the 15 newly independent higher education systems that developed rapidly between the mid-1950s and the late 1970s, after which economic problems caused enrolment to stagnate. Up to 5 percent of the age cohort enrolled in full-time higher education in the mid-1950s; this grew to about 11 percent in early 1960s, and to 12.5 percent in the late 1970s. Affirmative action played an important role in Soviet higher education policy. While in the late 1940s, less than 30 percent of higher education students studied part-time or by distance learning, by the early 1960s this figure exceeded 60 percent.[7] These were young industrial workers in particular who had entered *institutes*, ultimately to become engineers. Higher education was also one of the few legitimate means for young people to leave the collective farms and their intolerable living and working conditions. The total enrolment in higher education can be estimated to have been approximately 20 percent of the respective age cohort by the late 1970s. This tripling of enrolments in around two decades, from the mid-1950s to the late 1970s, would be difficult to repeat. Although enrolment has again been growing in the 1990s, the higher education system faces growing difficulties in facilitating not only relevant, but often perhaps any learning experience at all for the students. With all due respect to the elderly, even with their wisdom and life experience one still has to say that when the average age of the faculty climbs beyond 60 years, as is reportedly the case in Russia,[8] it may lack a certain critical vitality that is particularly important in introducing change.

Although numbers are often seen as having the power of ultimate arguments, statistics offered by various sources on Russian higher education vary greatly. Differences of up to 20 percent seem to be rather common, but sometimes they are much larger. The unreliability of East European statistics is one of the main reasons why this book offers no comparative tables. Instead, certain basic figures are introduced throughout the text in order to give an idea of the scale of Russian higher education. In the year 2000, it had reportedly 914 higher education institutions out of which 580 were run by the state, the rest being private or *non-state* higher education institutions as they are called in Russian. In 2000 the system accommodated 3.6 million students from a total population in Russia of approximately 148 million. According to some reports by 2002 this figure had reached more than 5 million students,[9] meaning that the gross enrolment rate had exceeded 40 percent. The total teaching staff in Russian higher education is 282,000 out of which 32,000 teach at private higher education institutions.[10] Unfortunately, there is no way of estimating how many individuals are involved in the 282,000 extant faculty positions. It is a known fact that many of the faculty hold multiple jobs and, therefore, the 32,000 private university teaching staff are most probably counted atleast twice, but that is in all probability all one can say about it. A recent OECD report estimates the faculty student ratio in Russia to be approximately 1 : 7,[11] but it is not easy to discern how such a figure has been reached. A better response is that currently nobody really knows what is going on in Russian higher education, and given that, this chapter is but one speculation, based on limited information, among the many.

Countries that fell under Soviet domination after the Second World War had had a university tradition long before the war, usually following the Humboldtian traditions that often provide a reference point for post state-socialist reforms. For several reasons, in most of the successors of the former Soviet Union—apart from the three Baltic States, which were incorporated in 1940—there is no such reference point to autonomous higher education. With Stalin's "Great Break" in 1928–29 and the purges in 1930s, the last remnants of liberal intellectual tradition in the Soviet Union were wiped away. The reform of higher education in the Russian Federation is thus far more problematic, but with its relative isolation and long history under the communist system it provides an interesting example to be compared with other countries in the region and perhaps even beyond. Some of the major recent developments in Russian higher education are examined below in the context of the 1992 *Law on Education* and its 1996 amendments,[12] various lower-level regulations such as the *State Educational Standard for Higher Professional Education*, and the author's personal experience of working on various projects in Russian higher education in the years 1996–2001.

Redefining the Mission for Higher Education

Higher education in Russia and many Central and East European countries has traditionally been quite rigid. Four to six years of full-time study led to a diploma in higher education. Although this diploma was often considered the equivalent of a master's degree and the nominal course-load generally exceeded that of similar courses in most European countries, teaching methods were not what one might consider as being the norm in contemporary universities. Thirty-five to 40 weekly contact hours, mainly lectures, constitute the very core of university teaching to this day. Opportunities for individual reading and writing, small-group discussions, and other forms of learning aimed at the development of critical thinking are limited. In Russian universities faculty remuneration is calculated according to the number of lecture hours provided, constituting a powerful source of faculty resistance to active teaching methods and independent student work being introduced.

Development of the skill of critical thinking was never the focus of communist higher education. Lenin had serious problems with the bourgeois cader in his universities who became at times perhaps more critical than necessary. An excerpt from his letter to Maxim Gorky characterizes his position well: "The intellectual forces of the workers and peasants are growing and getting stronger in their fight to overthrow the bourgeoisie and their accomplices, the educated classes, the lackeys of capital, who consider themselves the brains of the nation. In fact they are not its brain but its shit."[13] Lenin, and later Stalin made strenuous efforts to eliminate liberal thinking from Soviet universities. Like with every radical reform, the beginning of the Bolshevik reforms in Russian higher education was a difficult one. Communists constituted but a small minority among the faculty where the majority of positions were held by old bourgeois professors. The survival strategy of the latter was rather similar to that of the many of the post state-socialist university faculty some 80 years later—"Sometimes under the cloak of new names, the old professors managed to continue teaching their specialties."[14] And as it is now it was the same after the Russian revolution—students constituted the main change agent. Bolsheviks used the powerful student party cells to control what and how was taught in the universities. They seemed to have succeeded in the mid-1930s when academic ranks—abolished after the Revolution in 1917 along with military ranks and many others—were reestablished.[15] By then a new cader had been trained in the spirit of communist ideology and a new hierarchy could be established.

The role of the universities changed during the initial years of communist dictatorship. Idealism as the inspiration of the university was replaced by a strongly functional approach. The task of universities was no longer defined in terms of searching for truth, as this had already been revealed by Marx, Lenin, and Stalin, but rather the dissemination of the revealed truth and the preparation of the country's industrial–military complex for the ultimate battle against the imperialist world. Particularly Stalin's reforms in 1928–29—discontinuation of the New Economic Policy (NEP) and merging the state and party structures finalized the establishment of the totalitarian order. In this new order engineering schools assumed the function of producing the new elite. The nation's entire resources, intellectual and material, were mobilized for a single purpose. In studying the Soviet Union it is important to remember that it never did abandon the idea of the communist world revolution, although at different times the strategies to achieve that goal varied. During the hearings before the Subcommittee on Europe and the Middle East of the U.S. Congress in 1988, the role of the military in the Soviet system was revealed in its true sense: "There are several reasons why the military does consistently much better than the consumer goods sector—well, one, they have the pick of the resources. The best mental talent, the material inputs, the financial capital and so further, they have the pick of resources."[16] This is also confirmed by Sokolov and Tiazhel'nikova: "In the 'box' [closed organizations without a street address, only a P.O.Box—V.T.] institutes and enterprises worked the best researchers and technical designers. Their work was not only better paid but also carried high social prestige."[17] The idea of industrialization for primarily military purposes predominated the Soviet higher education until the late 1980s, defining the content, as well as the form of studies with the result that about 60 percent of students studied engineering and natural sciences while the majority of the remainder were being prepared as future teachers.[18] The Soviet Union was the first country to make higher education a part of its productive force. It had no need for the liberal ideas that the *bourgeois university* had traditionally carried. The Soviet Union only needed a higher education that would allow it to modernize its industry and armed forces as quickly and on as large a scale as possible. Needless to say that many similar ideas, serving the needs of industry, can be found in higher education policies in Western countries since the 1960s.

Many post-communist higher education systems are undergoing rapid diversification.[19] New types of institutions are mushrooming: private and profit-oriented, short-cycle, colleges of liberal arts, business schools, and so on. Restructuring higher education appears to be more complicated in the Russian Federation than in other East European countries, for several reasons. Even after the breakdown of the Soviet Union, the higher education system remains very large. Of the approximately 840 state higher education institutions operating in the Soviet Union in the mid-1980s, 566 remained within Russia's borders, with an average size of slightly more than 5,000 students. Over the past decade several new universities have also been established. In addition to these there are more than 300 private higher education institutions, all of them established since the fall of the state-socialist regime. This number, however, only includes those institutions that have been licensed by the Ministry of Education. In view of complicated licensing procedures in Russia and developments in other East European countries where many higher education institutions operate without official authorization, the number of active institutions in Russia may be much larger. But even if there is a strong demand for a higher education system that can react more swiftly to changing needs, the official policy seems to be to maintain the stability of the state sector; this seems to be leading to a wider gap between higher education and public expectations and is, in fact, creating a greater need for private initiatives. Attempts to tighten control over these initiatives pushes private higher education further toward the informal sector of the economy. Egorshin, for example presents discrimination against private higher education institutions as one of the main problems in Russian higher education.[20]

The need for Russia to transform the formal structure of higher education and to accommodate its new segments does not correspond with either of the options applied in other countries in the region, that is a uniform system consisting only of university-level institutions or a binary system of academic and vocational higher education institutions. The Russian Law on Education defines the whole higher education system as a system of *higher professional education* (Art. 9.5).[21] This could be interpreted as a return to a long tradition of polytechnic education in the Soviet Union, and it relates closely to the second of the three strategic areas of activity that were supposed to lead to communism: collectivization in agriculture, industrialization of production, and the Cultural Revolution. The intended shift of higher education from the academic to the vocational sector is encouraged by permitting graduates of secondary special (vocational) education institutions to obtain higher education qualifications more quickly than graduates of general secondary education institutions (Art. 24.3). Such a definition of higher education is somewhat unusual. Authors of a recent OECD report[22] seem to find it difficult to grasp that, at least formally, in Russia higher education is a direct continuation of the vocational education track instead of being a form of academic secondary education as is the case in most countries. The situation becomes even more confusing when one realizes that everyday practice is quite the opposite of the formal logic on which the educational system has been built. The answer to this puzzle can be found in the Soviet tradition of *useful* education in contrast with *useless* liberal education. In Russia, as it was in the Soviet Union, each graduate receives a professional qualification. For example a graduate from a philosophy department is awarded a professional qualification—*philosopher*. An individual such as the late Paul Feyerabend would be terribly excited about such a practice.[23]

To return, at least formally, to the initial vocational ideals of Soviet education is somewhat unusual, even considering the recent history of higher education in the Soviet Union. Despite the strong vocational inclination in the whole educational system, graduates of vocational schools had no direct access to higher education and only 5 percent of graduates from *Technicums*—institutions generally providing four-year professional secondary education courses—could apply to higher education institutions directly, unless they worked for two compulsory years after graduation in an appointed position. Graduates from *Technicums* were generally disadvantaged, essentially because of the relatively low level at which science and math were taught at these institutions.

This strong stress on vocational higher education is unique within the countries of Central and East Europe, where meeting international academic standards has a high priority. It also contradicts some of the aims of the Russian system. Entering the international community as an equal partner has long been a problem for the Soviet Union and its successors. Driven partly by the wish for political legitimacy and partly by the need to compete in European labor markets, Russian educational authorities have made demands for the recognition of Russian higher education qualifications. To overcome suspicion of courses traditionally overloaded with social science studies such as scientific communism, scientific atheism, Marxist–Leninist philosophy, and so on, special attention is paid to the academic recognition initiatives of the Council of Europe and UNESCO. In addition to following the latter's procedures through careful analysis of diplomas and their transcripts, the CIS has tried to produce its own certificates on international equivalence through bodies such as *Inkorvuz*, the Board of International Equivalence of Higher Education.

Attempts to present higher education, including doctoral studies, as primarily vocational in character conflict with efforts to achieve automatic international recognition for bachelor degrees, specialist diplomas, and candidate of sciences degrees from Russian professional higher education, as the equivalents of the bachelor, master, and Ph.D. degrees of Western universities. The real problem may go much deeper than the apparent controversy between the national and international aspects of higher education policy. The extreme distance between the institutions and the decision-makers, as well as 80 years of manifold reform, mean that formal policies have little effect in higher education institutions. The conflicting strategic aims of the system may have few implications for institutional operations. Institutions admit the number of students they are told to admit and provide them with courses as prescribed.

The currently fashionable issue in many European higher education systems, the relationship between university and industry, was prominent in the Soviet Union before anywhere else, but it could create only islands of excellence for the industrial–military complex set in a sea of economic stagnation. An extreme example of this relationship was the virtual abolition of full-time study between 1958 and 1965. Students had to work in industry for the first two years after their admission to university, taking only evening courses, and studying full-time only from the third year.[24] The results, however, were far from those desired. In 1988, "the USSR [made] as much money selling manufactured goods to the United States as the Ivory Coast,"[25] a fact that illustrates the problems of transforming the contributions of a large, technically oriented higher education system into industrial production. The separation of information from decision-making along with the delegation of responsibility to a level above criticism, may lead only to the decline of higher education and the sectors depending upon it.

The abolition of the mechanism of central planning has severed the artificial link between higher education and the labor market. The former Soviet system of planned admissions began with enterprises assessing the demand for graduates for their branch ministries. The aggregate figures were then passed to *Gosplan*, the State Planning Committee, which then ordered some 45 different federal or national ministries governing higher education institutions to produce a given number of graduates for any given year. This system has been partly dismantled. The new procedure, in which proposals circulate between the institutions and the Ministry of Education and eventually lead to admission quotas, is probably even less related to need than the earlier mechanism had been. Higher education institutions do not have the organization or facilities to form links with employers. Considering the direct relationship between student numbers and the scale of institutional funding (Art. 41.2), the institutions may find it difficult to fulfil ill-defined labor market needs rather than fighting for additional students and the funds that accompany them. The mechanism is clearly dysfunctional. Half of all higher education graduates in Russia cannot find a job in the professional field they are trained for,[26] and almost 20 percent of all engineering graduates are unemployed.[27]

The way the higher education system is labeled does not increase its relevance to the country's economy. If changing labels is supposed to stress usefulness and justify large public expenditure yet again, then this in itself hardly makes the system more responsive to developmental needs. In so far as the content of studies is static because of the extreme centralization of curriculum development and the lack of equipment for higher vocational training, drawing theoretically underqualified graduates of

Technicums into higher education may cause a further decline in the level of education provided, be it academic or vocational.

Failed Shortcut to International Recognition
The list of qualifications issued by higher education institutions has also been changed recently. The most important of the reforms so far is the abolition of the system of *unilevel*, mainly five-year higher education courses leading to a *diploma*. By the end of the academic year 1995–96 these should have been replaced by four-year bachelor degree courses, but because of cultural as well as legal obstacles, this aim has not been met. The new pattern is that the next educational level after a minimum of four years of bachelor studies is two years of specialist training leading to a master's degree, considered to be the equivalent of the former higher education diploma.[28] The three to four years of postgraduate studies leading to the Ph.D.-level *Candidate of Sciences* degree now requires six years of preliminary training instead of five, as it did before.

The addition of an educational level and subsequent 20 percent reduction in the time the average student spends in the system, may cut costs. However, there is also a further year of training for those wanting to start research degrees. Achieving compatibility with Western qualifications is another strong argument behind educational reform that contradicts local traditions, as well as the labor law—neither of which recognize the bachelor degree as comprising a complete component of higher education. A statement by the former chairman of *GOSKOMVUZ* (the State Committee of Higher Education—a government body that functioned as the Ministry of Higher Education until 1996 when it was merged with the Ministry of General and Professional Education, which was then renamed as the Ministry of Education in 1999) reveals the underlying intentions: "It is very important also that through state educational standards Russian higher education enters the world cultural and educational stage, and receives the necessary legal basis for international recognition of documents on education [i.e. degrees and qualifications] and the organisation of academic and scientific exchange and contacts."[29] Although the bachelor degree was introduced in the Law on Education, labor legislation has not been amended accordingly. The final result of this major reform is that students wanting qualifications recognized by employers and the State Committee on Labor must study for six years instead of five. Introducing another level after a relatively extended bachelor studies and preceding a relatively long period of doctoral studies may in fact diminish interest in a low-paid university career and encourage the already strong tendency to employ an ageing faculty. Although a difficult situation in the labor market alongside small state grants for students may force them to stay at university for as long as possible.

To put it simply, by adopting *The Higher Education Act* in 1992 the Russian Federation tried to take a shortcut to international recognition of its degrees and qualifications by replacing the entire degree structure. Although this might have worked with the international community, the first to fail to recognize the new bachelor degree were Russian employers, as well as some major universities who declined to award degrees to students who had, in their eyes, dropped out before the final year of study. Because the four-year bachelor degree in Russia lacks tradition, it became a sign of failure that not many people wished to hold. The entire reform was reversed in 1998 when the Law was amended to the effect that universities were allowed to award both old and new degrees. This was perhaps the clearest indicator that the most radical reform that Russia had tried to accomplish in the 1990s had failed. Russia's joining the *Bologna Process* in September 2003 has raised the issue again, problems, though, remain the same.

The Legal Status of Higher Education Institutions

Altering the legal status of higher education institutions is a core element of higher education reform in Eastern Europe. Changing communist legislation has an important symbolic value. Although every legal system allows a significant level of variant interpretation there is also an *old wine-skin new wine* conflict, which suggests that to bring about change, instead of reinterpreting old laws, new laws should be written. At the same time, it is hard to change a legal culture in which anything that is not clearly permitted in law has been automatically prohibited as is the case in the Soviet tradition. Such a culture requires extremely detailed legislation. New East European higher education legislation is largely aimed at drawing a distinct line between higher education institutions and the state, the latter having enjoyed unlimited control over activities within them through the agencies of executive power, law enforcement, state security, and units of the communist/socialist parties.

Legal reform is expected to lead to increased academic autonomy, as well as to better utilization of limited resources, thus providing institutions with additional opportunities to benefit directly from a system that is more rational than the one in which they were acting as extensions of the state. Despite the apparent difficulty of changing the legal environment and distancing academic institutions from political power, university autonomy seems to be the legitimate aim of such reforms in many countries—although knowledge of how to accomplish it is scant and academics are also often reluctant to accept additional responsibility for the expected cuts in posts and growth in competition. In this, however, a conflict emerges: while the state would happily surrender its funding responsibilities, it is still keen on controlling the academic content of higher education studies. For institutions the ideal would be quite the reverse.

A totalitarian *state* by definition does not recognize other entities in its territory that might have a will differing from that of *the people's*. In this sense one cannot speak of different legal entities in the Soviet Union, but only of various units of the state apparatus endowed with limited operational authority and lacking strategic decision-making powers. The status of the universities and other higher education institutions was similar. They were described as *legal persons* in various normative documents, but could not make decisions on curricula, financial matters, faculty recruitment, or student admission procedures. Universities had their stamps, letterheads, and the authority to withdraw money from the state bank to make the payments predetermined by the state authorities. Following the same line, curriculum development in higher education only entailed scheduling the weekly classes following a predetermined number of lectures covering predetermined issues in any given subject.

For several years, many East European universities, which had been forced to follow the Soviet line, have been claiming a new legal status to enable them to act as independent legal entities. Again, since the tradition and culture of the *collegium* is lacking, the only way to achieve this aim is to draft new legislation redistributing authority among the state, institutions, and units. Redefining legal landscapes, abandoning old vocabulary, and introducing the concept of limited state and legal persons in public and private law, are helpful in defining the expected changes in relations between the central government and higher education institutions.[30] The devolution of control over university teaching and research is a necessary, though not a sufficient condition for the development of academic freedom, whether an institution is owned by the state or a private body; however, this cannot be achieved without a significant increase in funding, which would, of course, also be easier to achieve within an internally consistent framework of legal regulations.

A major problem for the Russian higher education system as defined by the Law on Education is the limited awareness of other, radically different systems. The post-communist definition of the legal status of higher education institutions in Russia as legal persons carries a heavy load of historical continuity, and extrapolates the existing institutional structure and curriculum into the future. Defining higher education as a federal prerogative (Art. 28.23) appears to concentrate many of the functions traditionally exercised at institutional or unit level in the Ministry of Education. Further problems stem from the civil code, the logic of which does not allow the creation of public institutions that are not directly owned by the state.[31] The level of control and discipline the state expects to see applied in its institutions makes it very difficult to develop a public sector outside of direct state control without making it explicitly private and consequently complicating the issue of its funding by the state.

The Prospects for Private Higher Education

As a formal principle, the Law on Education stipulates that virtually any organization or individual can establish a higher education institution in the Russian Federation, something that was unthinkable in the Soviet Union. The logic of this is, however, highly state-centerd. Several of its articles reveal the apparent difficulty the legislators had in extending the traditional state-centerd regulations to other bodies involved in the establishment of such institutions. For example Article 41.9, according to which the founder cannot reduce financial support to an institution that raises additional funds, has meaning only in so far as the state is considered to be the founder.

For a privately founded and financed institution that is partially or entirely financed from the income it generates, Article 41.9 opens bizarre perspectives for the relationship between the institution and its founder. The founding person or body seems obliged to imitate the Ministry of Education, producing admission numbers and providing admission-based financial allocations, while the institution retains the right to offer additional services without reporting them to the founder. Legal solutions intended to limit the opportunities that state higher education institutions have to generate additional income on the margins of the centrally controlled system lose their meaning when applied to institutions based on private initiative. Applying traditional legal provisions to such institutions is difficult and makes their legally valid existence within a given environment somewhat problematic.

In so far as higher education remains highly state-centerd, the *state* does have a crucial role in the Russian system. However, as in principle it no longer directly rules the legal entities involved in higher education it is confusing to use the term without precisely defining it. With the new procedures, especially those related to state accreditation, the concept of a state higher education institution, which until recently was the only genuine form of such an institution, has suddenly assumed two differing and slightly contradictory meanings. The first meaning of the term stems from the fact that a state higher education institution is founded, owned, and run by the state and is largely or totally funded from the public budget. In the former Soviet Union this applied to all higher education institutions. The second interpretation stems from the Law on Education's stipulation of how this status is to be achieved via the accreditation procedure. If the state is concerned neither with foundation nor ownership, but is concerned with an institution's status and its authority to issue state-recognized degrees and qualifications, by implication a privately established and funded university could be called a state university. Through the state accreditation process, a higher education institution becomes a state institution, gains public funds for courses adopting state educational standards and, since the content and form of instruction are prescribed in detail, loses influence over the content of accredited courses. Courses not meeting the prescribed standards, however, will not receive public funding nor will its graduates receive official state certificates.

Since the Soviet Union disintegrated, the first meaning of the term has changed in practice in all 15 of the former republics. For example in Estonia universities with that status no longer exist. University-level institutions are divided into public universities (founded by the state) and private universities (founded by private persons or companies), both of which enjoy extensive formal and even greater actual autonomy in programmatic, as well as financial matters. There are also some vocational higher educational institutions with only limited autonomy. These are divided into state-owned institutions supervised directly by the Ministry of Education, and private vocational higher education institutions whose activities are little regulated. While this pattern needs to be further elaborated upon, clear differences between the institutional groups are emerging.[32]

The motivations underlying the Russian Federation's higher education policy remain largely implicit and current legislation makes it almost impossible to distinguish between institutional groupings. As higher education issues are in the hands of the federal executive, the stipulation in the Law on Education that the general division of all educational institutions into state and municipal bodies (Art. 5.3) cannot be applied to higher education institutions. The fact that the Law also defines them as legal persons does not alter their traditional treatment as extensions of the state apparatus. But the attempt to treat the recently established independent higher education institutions in a similar way is new.

Both interpretations of the term *state higher education institution* appear controversial and the prospect of independent institutions gaining state status and public funding through accreditation seems theoretical at a time of scarce resources.

The opposite may just be the case. Some marginal state higher education institutions have recently lost their accreditation and, in consequence, their state funding and the right to award state diplomas. In this way these have perhaps become private institutions. But this is a strange way to privatize universities, indeed. The link between accreditation and public funding is an additional obstacle for institutions seeking accreditation, since it is not just educational quality that is considered but also, usually, the limited availability of public funds. The current system holds out the possibility of accreditation for an insignificant proportion of approximately four hundred licensed independent higher education institutions. At present, not one of them is state accredited, although the objective was to accredit all higher education institutions within a year. Many East European countries have demonstrated that this is impossible if it is to be anything more than automatic accreditation of the traditional state-established institutions.[33]

Despite modifications to the system regulating higher education, neither significant change in institutional status nor a redistribution of power has occurred in the Russian Federation. Independent institutions cannot compete with state institutions as no proper mechanism for setting the true value of their qualifications has been set up, and the current regulations force them to copy the courses offered by traditional institutions. In this sense the situation resembles that of Romania, where new private institutions were initially forced to offer similar courses to those in existence before the December 1989 revolution and then later condemned for copying these courses and for their inability to establish new ones.

Academic Freedom and Quality Assurance
Russia has adopted a three-step process through which the quality of *higher professional education* is to be assured: attestation, licensing, and accreditation. In addition to these procedures defined in the Law on Education there is also a range of legally binding lower-level documents that regulate matters related to administrative procedures, financing, and educational process. The latter regulations have long traditions in the Soviet administrative practices and, in higher education, they take the form of standard statutes for institutions, course programs, teaching plans, and so on. Such a document—"polozhenie"—a university statute was first adopted in communist Russia as early as in 1921.[34]

A higher education institution that is uninterested either in its graduates' eligibility for the professions requiring a Russian Federation higher education degree (Art. 27.4), or in funds from the public budget, can avoid intensive contact with the educational authorities and teach whatever subjects it chooses—apart from military disciplines (Art. 11.2). However, state accreditation opens the prospect of fitting its graduates for regulated professions and of receiving public funds for itself.

State accreditation of a higher education institution is the third and final step in its foundation process. It is preceded by licensing and attestation. A license is issued by The Ministry of Education after examining the general material conditions of the institution, in particular its physical equipment; the license gives it the right to operate as a higher education institution, but does not imply state recognition of degrees and qualifications issued. Attestation is an examination process carried out by the semiindependent State Attestation Services, through which graduates' level of training is assessed against official educational standards in the form of state examinations; attestation is a necessary prerequisite for accreditation. Legally, accreditation means issuance of the state accreditation certificate to an institution that has passed attestation. Since higher education is considered a federal prerogative, all the related formalities and paperwork are done by the Ministry of Education in Moscow. Given that Russia is geographically a vast country, the Higher Education Sector of the Ministry of Education is divided into four geographical divisions: for the Central European area of the Russian Federation, for the North and South European region, for the Ural and Volga region, and for Siberia and the Far East. Here, one should bear in mind that the meaning of the notions like Central Europe or Southern Europe in the context of governance of the Russian Federation are in no way related to the use of the same terms in Western political science. Each of these four geographical divisions is actually larger than the whole of Western Europe.

By law, legal entities of all types, as well as individuals, regardless of country of origin, can act as founders of nonmilitary higher education institutions in the Russian Federation (Art. 11.11) and apply for a license. A newly established institution can start providing educational services from the moment it is licensed by the federal executive agent (Art. 28.20), in this case the Ministry of Education. To register and license an institution, the founder must submit an application form and other documents, the most important being the charter of the institution (Art. 33.3). The founder is responsible for compiling the charter and may appear to have extensive freedom to define the scope of institutional activities and management. However, the charter has to conform to the standard statute for this type of educational institution, approved by the federal government (Art. 12.5) and providing the federal executive with unlimited power to regulate institutional activities.

Before the license is issued and the institution can start to function, it is examined by an expert commission appointed by the Ministry of Education. The physical equipment must fulfil particular conditions, for example meet construction and environmental health protection standards. While the content of studies is not checked at this stage (Art. 33.2), the qualifications and numbers of teachers are considered (Art. 33.9). This puts certain obstacles in the way of establishing a new higher education institution. It is a vicious circle: to commence tuition an institution needs to be licensed; to become licensed staff have to be recruited; and to recruit staff an income is needed. However, most of these institutions do not have any other income than tuition fees.

In order to establish, for example a small independent college in Novosibirsk, all the paperwork has to be done in Moscow. Concentrating the licensing function at the center places bureaucratic obstacles in the way of establishing independent higher education institutions. It also leads, despite the capacity of the federal power to control the country, to a massive grey area in higher education that in particular threatens the rights of students to valid credentials in return for their commitment and outlay. There are perhaps a number of higher education institutions that choose not to seek formal status as higher education institutions and operate with no guarantee to the quality of the service they provide.

Having a license does not mean that a particular institution's graduates are equal to graduates of accredited institutions when applying for jobs requiring a higher education qualification (Art. 27.4). Degrees and qualifications are validated through state accreditation. The graduates of nonaccredited institutions can also obtain a valid state higher education certificate by being examined as external students at accredited institutions.

State accreditation, as defined by the Law on Education, is a twofold process. The necessary prerequisite is attestation, which, as mentioned, assesses the education of graduates in the light of state educational standards. Attestation is carried out by the centralized state attestation services and has to be renewed once every five years (Art. 33.19). At least 50 percent of an institution's graduates need to pass the required examinations (Art. 33.20). A new institution can apply for attestation not earlier than three years after receiving a license (Art. 33.20). Since the results of three consecutive cohorts of graduates are taken into account, attestation cannot be confirmed less than five years after receiving the license.

After attestation, that is after establishing that the quality of graduates reaches state educational standards—which, incidentally, does not determine the output in terms of students' knowledge and skills but the input in terms of the subjects taught and respective loads in hours—the Ministry of Education issues an accreditation certificate listing the study programs covered by accreditation within the institution. This endows the institution with the right to issue state certificates for qualifications and demonstrates entitlement to allocations from the public budget.

State accreditation may be withdrawn if attestation is not renewed within five years, with subsequent loss of official status, the right to issue a certificate bearing the arms of the Russian Federation and of public funds related to student numbers (Art. 33.17 and 33.20). It can also be withdrawn on the basis of the repetition within two years of a complaint by the state attestation services about the quality of the education provided. The complaint can be based on an application approved by a general meeting of the students, or on the recommendation of the state employment services.

The former situation leads to sanctions if less than 50 percent of eligible students pass the state examinations. The latter procedure includes an opportunity for students to query the status of their own university; at a general meeting they can question their institution's state accreditation and in doing so voluntarily invalidate their own qualifications. The probability of this happening cannot be high, but it could be an additional weapon in the hands of strong students' organizations, like the student party cells were in the 1920s.

How the state accreditation mechanism controls the content of studies has less to do with assessment against a general standard than with facilitating a powerful procedure for imposing centrally established curricula. In the Soviet Union, the content of courses was completely predetermined by the central authority. The new regulations give institutions marginal powers to change the total amount of teaching hours devoted to each subject area (not more than 5 percent), and to add an extra 10 per cent of specialist subjects to the total size of a course. Elective subjects for students can form up to 5 percent of the whole course.

The activities of higher education institutions are controlled in detail through minor normative documents of the federal government or Ministry of Education. The documents deal with the general running of institutions, for example the standard statute adopted as their charters or they regulate teaching activities as exemplary course programs, subject programs, and so on. They are not intended as guidance, but are binding on the institutions offering accredited courses. Ignoring these regulations represents violation of state educational standards incurring loss of status and funds. The Law on Education does not specifically address the role of these documents, but certain articles reveal their intent. For example Article 28.5 stipulates that federal approval is needed for the appointment of the head of an educational institution, unless a different procedure is defined by the Law on Education or the Standard Statute of a particular type of institution, which clearly demonstrates that the force of the Statute on Standards is equal to that of federal law.

Most of the regulations determining the content of studies are related to the state standard for higher professional education and together form a large set of documents. The standard consists of three major parts: State Educational Standard for Higher Professional Education in Terms of General Requirements; State Educational Standard for Higher Professional Education—Classifier of Fields of Studies and Specialities of Higher Professional Education; State Educational Standard for Higher Professional Education Requirements for the Minimum Content and Level of Training of Graduates.

The first document defines the general framework for higher education studies in the Russian Federation, the nominal duration of studies at each level, and the maximum workload of students, which cannot exceed 54 hours a week, out of which half is to be directly supervised by a faculty member. The second document concerns study directives and courses, providing a complete list of programs and courses offered in the country as a whole. A new course must conform to this list; for example in humanities and socioeconomic sciences, the standard lists 23 fields of studies and 73 study programs (*spetsialnost*). The third document provides a detailed list of subjects taught under every program and the teaching time allocated to each subject.

Federal Determination of the Content of Studies
The level of predetermination of the content of studies is also interesting. To receive a diploma, a student following program *071900: Information Systems in Economics*[35] has to spend a certain amount of time on certain subjects, of which half involves contact time with a faculty member. In the five-year program that amounts to 8,316 hours of study: 1,690 hours are spent on humanities and socioeconomic subjects (philosophy, a foreign language, cultural studies, history, law, sociology, political science, psychology, and pedagogy and physical education); 1,200 hours on mathematics and general natural sciences (mathematics and informatics, theories on contemporary worldview); 2,170 hours on professional subjects (economic theory, history of economic thought, economics and sociology of work, finance and banking, computers and computer networks, fundamentals of algorithms and programming languages, theory of information systems in economics, information in economics, labor safety regulations); 2,000 hours on specialist subjects (data and databases, development and application of computing programs in economics, design of information systems in economics, economics of informatics, theory of optimal conduct of economic systems, modeling of economic processes, management, marketing, bookkeeping, statistics, technical-economic analysis of business

activities, fundamentals of business). An institution can add 806 hours of specialist subjects, and student's electives equal 450 hours.

The standard for a new course, bachelor course *522400: Religious Studies*, specifies that students take: 714 hours of mathematics and natural sciences, 1,802 hours of general humanities, social, and economic disciplines consisting of political science, economics, as well as 408 hours of physical education. According to the standard, the bachelor degree-holder in religious studies "shall master the nature of man and the meaning of existence," have a command of economics and natural sciences, be computer-literate and be good at sports! However, the 12-page standard omits to mention the word God. One can only presume that a similar program was taught for decades by the very same professors under the title *Scientific Atheism*.

The standard reveals two characteristics. First, students do not seem to spend more than one-third of their time on subjects related to the profession they wish to join—a fact questioning the definition of higher education as primarily professional in the Russian Federation. Second, if an independent higher education institution wants to introduce an economics program, it has to teach all these same subjects in order to have its graduate qualifications validated. This means that although the so-called non-state higher education institutions are funded from private sources and possibly governed as private organizations, the content of their studies is determined by the state.

The Financing of Russian Higher Education

In following the Law on Education, the Russian Federation seems to be continuing the practices of an unprecedentedly large public higher education system for a country at its current level of economic development—a country where "36 percent of the population, or 52 million people live below the subsistence level, set at a dollar a day."[36] The Russian Federation is committed to covering higher education costs from the public budget for at least 170 students per 10,000 inhabitants, that is approximately 25 percent of the age cohort, a total of about 2.55 million students. At least 3 percent of the federal budget is to be allocated for higher education (Art. 40.2).

Maintaining the existing number of publicly supported students is made possible by keeping *per capita* teaching costs at a very low level through a policy of zero investment and low salaries. Paying faculty so little has also allowed Russia to maintain a rather generous faculty student ratio of around 12:1 and set a target of 10:1.[37] While investments have been very limited, the Law on Education still stipulates salary levels for faculty that are higher than elsewhere in the region. They have to be kept at the level of at least twice the average salary in Russian industry (Art. 54.3); the salaries of the other employees in the system are kept at the level of equivalent positions in industry (Art. 54.3).

Recent history demonstrates the difficulty of fulfilling similar promises and allocating planned funds: "Higher educational institutes obtained equipment worth 55.2 million roubles (instead of the planned 71.1 million) in 1991, with only 22 million roubles from the state budget (it should have been 32 million roubles). In 1992, higher educational institutes received only 14 per cent of the capital investment funding they require."[38] In the year 2000 the Russian government was able to spend 10,875 roubles (US$360) per higher education student, while the funding norm it had established was 63,800 roubles (US$2,120) per student per annum.[39] The fact that the Russian government has been able to meet only one-sixth of its funding commitment to higher education offers the key to the recent rapid growth, if not explosion in the student enrolment. In a desperate attempt to balance their budgets, the Russian universities have been using their gradually growing administrative freedom to aggressively recruit fee-paying students. While some Russian commentators like Alferov and Sadovnichi present it in terms of a growing social value of higher education,[40] there are some suggestions to the effect that more than anything, the market success of Russian higher education is being fed by the war in Chechnya that drives young men to universities in a hope that by the time they graduate and their postponement of the army conscription expires, the war will finally end.

The economic situation in Russian universities looks difficult indeed. It is particularly so in provinces that receive less than 10 percent of the total Russian higher education funds, the rest being shared between St. Petersburg and Moscow, the latter being by far the greatest beneficiary. Although the law sets the faculty salary as described earlier, what is actually paid is considerably less than that. The budgeted higher education faculty salary in the Sverdlovsk region in 1997 was only 80 percent of the average salary paid in that region, compared to 130 percent of the average they received in 1987.[41] A further problem is that even this modest remuneration is not regularly paid. For example in the Sverdlovsk region as of January 1, 1999 academics had not received their salaries for an average of 7.1 months.[42]

Frequent declarations of *la morte de la science russe*[43] [the death of Russian science] stand deeply incongruent with the official federal policy. The monthly salaries of researchers are in the range of US$100 and some have not been paid for months. Russian science may, indeed, be dying. For example in the years 1991–97 the number of researchers in Russia was cut by 50 percent,[44] not as a result of government policy to reduce the research sector, but because of people leaving. As there is no policy on how to deal with the issue, a significant part of the research sector is drifting toward its end and a large number of formerly high-status academics are suffering. Having spent decades in research many of them lack the necessary skills and attitudes to move to private business. They continue to go to their offices and research labs daily even if they receive no pay for it. The funding for experimental work is no longer available, so the researchers continue working with theories on paper.[45] In her recent study, Petrova describes the survival strategies of researchers and their silent rebellion against the brave new market-driven world—shopping cheaply in the marketplace instead of shops, withdrawing into their *dacha's*, turning off the radio and television, and hoping that this is a bad dream that sooner or later will end.[46] Then, perhaps, their social status and benefits would be restored. One of her respondents, a 55-year-old male describes how these changes have altered his style of dress:

> ...to keep up, [my] psyche has changed. Say for example, to maintain my image and worth as a human being..., I have even started to dress differently...—very democratically—pullover and jeans. In the earlier days this was the way to rest from daily work, but now this has become my second skin.... Even psychologically—I can't [dress] like "the new Russians"—everybody wearing suits, neckties. With this they have occupied a certain niche. For me personally it is uncomfortable to follow fashion and

dress like a bank employee. Maybe that would have reflected the status 'til the time everything changed, but now it's not my attire, not my form.[47]

The formal commitments on funding may be meant to reassure the academic community about the federal administration's intention to support it and also to safeguard its own position against angry professors on the streets. Recently, the Russian government has promised even higher recompense to academic staff in higher education, equal to three average salaries in Russian industry,[48] but without being able to deliver it. While, according to its very principles, in Russia the state runs higher education and, according to its constitutional law, higher education is provided free, it is currently estimated that approximately 80 percent of Russian higher education funding comes from private sources. This is particularly paid by contract (i.e. fee-paying) students who, according to formal regulations, can constitute no more than 20 percent of the total student body, but the actual number of which can be estimated at the level close to 60 percent of the total student population. A very rough estimate would give a figure at around two-and-a-half million fee-paying students in Russia who actually keep the whole system running. Following the funding source, Russian higher education may currently have one of the largest shares of private funding in the world, although the system is formally public.

Russian legislators have created a loophole in the Law on Education that allows higher education institutions to generate additional funds through the following activities: (1) renting out property (Art. 39.11), funds so generated can be used to support and develop educational process; (2) accept voluntary contributions from Russian and foreign organizations and citizens (Art. 41.8); (3) admit fee-paying students—in addition to those admitted on the state quota, although paying students cannot exceed 25 percent of the state quota in fields like law, economics, state and local government (foreign students do not come under the state quota, and institutions are free to admit as many such students as they can); (4) offer additional paid educational services; (5) engage in entrepreneurial activities as defined in the institution's charter. Additional funds cannot be a reason for decreasing funds from the founder—here it seems quite clear that the *founder* means the state (Art. 41.9).

Higher education institutions would seem to be in a good position to secure their income. However, there are serious limitations to this set by other laws such as the Law on Taxation and the Law on the Budget. These laws greatly restrict the activities encouraged by the Law on Education: any income generated is heavily taxed or, if noticed, is immediately transferred to the state budget. The approach taken under conditions of conflicting federal laws depends, in any particular instance, on the decision of the tax authorities that are under permanent pressure to collect more rather than less tax.

As the regulations on how to account for this money remain contradictory, a large proportion of the funds circulates as completely unaccounted cash, or is moved through organizations other than the university where a particular student studies. It may be a private university, run in parallel with the state university by the state university's leadership, but it may also be a business organization or foundation. In addition to the fact that universities are vitally interested in retaining the fee-paying students and tend to apply different academic standards to them, such a use of fee-paying students makes higher education prone to a variety of corrupt practices.

In addition to higher education institutions' generating a major share of their funds through charging fees to students, a practice that threatens the integrity of the entire institution of higher learning in Russia because of the manner in which it is done, individual faculty members have also become entrepreneurial and create even more risks for their universities. Although faculty pay remains modest there is a vast market for private tuition available in Russia that allows university faculty to generate additional income for themselves. For the time being, competition for state-subsidized student places in universities remains extremely high. It is a common practice that students aspiring to occupy such nonfee-paying positions take a private tutor during their last secondary school years, at a significant cost. The tutor, who is normally a faculty member of the university the student intends to apply to, helps the student to prepare for the entrance examinations. Not only that but the tutors also often assist their students in passing the test successfully, making sure that a student receives full value for their money. This has created a rather bizarre situation when the strongest voices against the introduction of the state-administered final secondary school exams that would also count as university entrance exams (such as, e.g. the French *baccalaureate*)—a step that would ease the situation of both the students and faculty as well as help to introduce common standards in Russian education—come from the camp of the university faculty; a group that is vitally interested in maintaining the market of tutorial services that allows them to survive in difficult times.

Conclusions

Reforming a higher education system is always a difficult task and there are not many successful examples at hand. It is, however, more difficult in a country like Russia, which is not ready to admit that its higher education needs reform to restore its intellectual integrity and social and economic relevance. Although many things are changing at the institutional level—often in ways different to those expected by the federal government—the latter still tries to conduct the system as if it had the information it needed to make critical decisions, as if it had the freedom to make the necessary decisions, and as if these decisions were implemented at lower levels. Unfortunately, none of these assumptions seems to be valid. The Ministry of Education in Moscow does not have basic information on the performance of the higher education system, and even if it had, under current social pressure it would not be able to make rational decisions; however, even if it had made the decisions the current management structure would not allow it to follow up the extent to which decisions could be implemented.

There is a discrepancy between the way Russian higher education appears from outside and how ministerial officials perceive it. The following is a quote from the State Education Standard: "Today, higher education institutions are free from ideological and administrative regulation and supervision, they have gained real independence and all the possibilities for the realisation of academic freedoms in accordance with the requirements—freedom of teaching, freedom of research, and freedom of training."[49]

Inaccurate interpretations of reality like the one just mentioned are not rare in East European higher education.[50] Perhaps producing them is a part of the daily work of politicians. However, if academics cannot reject such political statements or, even worse, are actually interested in distorting the picture still further, something is fundamentally wrong with their moral principles. A university without a moral commitment to truth is not a university.

The problem may well be that the problem itself has not been understood. The most remarkable development in higher education since the breakdown of the Soviet Union is the introduction of four-year bachelor courses. In financial terms, this step could help to save 20 percent of the teaching cost per student. It could also help to gain recognition for Russian qualifications abroad. But recognition will be hard to come by as long as the main problem related to curricula in Soviet and Russian higher education, their overload with general subjects—often just modifications of the old *red* subjects and taught at less than the optimum level—remains unresolved. While aiming at international compatibility and recognition, the most serious obstacles, local traditions and lack of support from other federal agencies, are being neglected.

Many provisions of the recently adopted Law on Education reflect over-politicization of educational issues. Because of the heavy stress by the former Soviet leadership on quantitative targets in education, the Russian legislative and executive powers do not seem to have had a chance to give up *pure socialist output.*[51] Just as TV sets were once produced for warehouses, higher education graduates now swell the ranks of the unemployed. In current economic conditions and in terms of the efficiency of the system, this may lead to considerable reductions in teaching staff, especially those teaching general subjects who often had long careers teaching social sciences in Soviet universities.[52] In some of the regions of the Russian Federation, radical cuts in academic staff can take place under the recent initiative to merge higher education institutions in particular locations into large universities, an initiative that certainly does not find great support among the faculty. The state-supported share of higher education may also have to be reduced since maintaining it is increasingly expensive. Re-labeling higher education as *higher professional education* may not help to justify its heavy burden on the public budget for much longer.

It may be that in addition to satisfying the public demand for the communist principle of "to each according to their need," both the basic guidelines of the current Russian higher education legislation—protection of the state sector from competition, and unrealistic commitments to student numbers and financial support—are driven by the powerful university community. In this respect, cosmetic changes in terms of describing higher education as professionally useful serve to make the expenditure involved more acceptable. But in very real terms there is almost no cost left for the state to carry anyway. In the guise of public provision of higher education, Russian higher education has over a decade become more private than any other higher education system in the world. Ignoring this harsh reality further only generates further corruption.

Chapter 5
Market as Metaphor in East European Higher Education[1]

The success of post-communist reforms in Eastern Europe is often measured by how far market-based regulatory mechanisms have replaced previous administrative central planning procedures. There seems to be competition to have the most extensive market reforms. Poland is famous for its "shock therapy," which liberalized the prices of basic goods overnight. Estonia is praised for its monetary policy. The fame gained through far-reaching reforms, the social cost of which is often ignored, is seen as political capital leading toward EU membership, which in itself may well be a substitute for another recently lost, utopia. Other, less successful countries still proclaiming the same political goal may, in the long term, need to consider less attractive alternatives for international cooperation. Developing markets have a strong impact on the public sector, including higher education. While European states are decreasing their support to higher education and making space for the market, the share that markets are demanding in East European higher education is unprecedentedly large. Hundreds of recently established private universities operate in an ill-regulated environment, and conditions of widespread moral relativism question the values that academia had succeeded in preserving under a hostile political regime. Neave has interpreted recent changes and rising managerialism in European higher education in terms of the American dream becoming a European nightmare,[2] a rather mild turn of phrase to describe the consequences of marketization in East European higher education.

In his seminal work, Clark defines the higher education system of the (former) Soviet Union as one of relatively limited market coordination.[3] This opinion can be considered a gross exaggeration just as long as one does not interpret the interaction of the State Committee of Security (KGB), Communist Party structures, and the so-called trade unions guarding the ideological soundness of higher education in terms of the expression of institutional markets. The latter approach cannot be fully excluded and Clark seems to imply a similar interpretation. However, in adopting it, one seems to be proposing that any social interaction is an expression of a market. Again, this view can be justified, but only in certain contexts where *market* is applied as a metaphor. If not, the strivings of East Europeans over decades are denied meaning.

As the term *market* has become very much value-laden it is not easy to distinguish between its two uses, the metaphoric and the literal. Having more markets, whatever this may mean, and farther-reaching market reforms is considered to be intrinsically good, whilst having no or fewer markets is bad or worse. Following this logic things have got better: under communism there were few or no markets, now we have lots of them. Still, one may not fully agree with this. The role the markets play in various sectors differs greatly. Even within higher education in some parts markets have almost no role to play, while in other places the role of largely unregulated markets threatens the principles of academic integrity in addition to violating the principles of consumer protection. How else can one define the operations of any office selling worthless degree certificates instead of providing higher education at the international quality level that it claims it is doing?

It may well be the case that economic difficulties and the dissolution of the central redistribution and planning mechanisms have caused major difficulties for state-run higher education. Inadequate funding does not allow the undertaking of reforms that reach beyond the cosmetic. Changing the structure of the demand for higher education finds its responses out of the traditional public sector in the large number of new private universities. Their extreme responsiveness, however, not only threatens the principles of academic integrity but may also lead to direct mistreatment of the students as customers by low-caliber universities and diploma mills.[4]

Reforms in Higher Education

Today's higher education in the East European region differs considerably from that of a decade ago. One of the most remarkable changes is the virtual disappearance of the largest group of institutions in higher education, the specialized *institutes*. Instead of the medical institutes, pedagogic institutes, and polytechnic institutes that once constituted the vast majority of institutions, nowadays, alongside the limited number of state universities, one can find pedagogic universities, technical universities, and medical universities. The extent to which the alterations in their titles reflect a deeper revision of institutional missions and teaching/research practices remains open to question. One could argue that the changes reflect academia's denial of the role attributed to it by the totalitarian political regimes in the process reproducing the communist *nomenklatura* and the desire to distance itself from some of the personality-molding activities previously practiced under the supervision of the Communist Party and the national security structures. However, beyond the point that the latter structures have been dissolved or deeply reformed, the names of the institutions may well exemplify the limits of radical reforms of public higher education in given political and economic conditions.[5]

Comparing the total number of higher education institutions in 1988 and 1998 one can see significant growth—from 50 percent to 300 percent and even more, depending on the country. The exact number of higher education providers remains unknown because many of the recently established universities operate beyond the limits of the recognized higher education sector, relying heavily on neoliberal market principles. The latter may, in the worst cases, mean a simple exchange, cash for a degree.

The Russian Federation clearly represents the quantitative development of the private higher education sector in the region. On the one hand, its traditionally high enrolment in higher education, which reached 20 percent of the respective age cohort in the 1960s, has prevented an explosion in Russia like that in post-Ceauçescu

Romania where the number of higher education providers tripled in a matter of months.[6] On the other hand, large and highly centralized Russian public higher education has been unable to meet the training needs of emerging private businesses and banking, the need for an explicitly élitist educational environment, as well as the training needs of those whose political inclinations or academic qualifications did not allow access to higher education earlier. Moreover, the majority of the former Republics of the Soviet Union including, rather surprisingly, some of the traditionally rebellious Baltic States[7] represent similar trends and are adopting quite similar policies. These two factors in interplay have defined relatively balanced conditions for the development of private higher education in Russia.

Still, the quantitative changes are remarkable. After the formal breakup of the Soviet Union in 1991, out of its 840 higher education institutions 566 fell within the borders of the Russian Federation. In addition to these institutions, the federal educational authorities sanctioned the activities of 334 private higher education institutions by 2000. Between 1991 and 1993 in Moscow alone 33 private higher education institutions were established.[8] Considering the bureaucratized and highly centralized procedure of licensing higher education institutions by the Moscow-based federal educational authorities, one can but speculate about the number of institutions operating without proper authorization. It may not be an exaggeration to presume that at least the same number of unauthorized institutions are operating in various forms. Countries like Estonia that initially licensed all new higher education providers ended up with 48 instead of the former six or Romania where one-third of the new universities were licensed, hint at the proportions between legally and illegally operating higher education providers. Summing up all the higher education institutions established in the region over the decade, one easily ends up with a thousand or more institutions established to meet various needs. They represent a wide range of academic excellence and with student enrolment ranging from as low as the Theological Academy in Tartu, Estonia with its 25 students[9] to as many as 6,500 in the Belarus Commercial University of Management in Minsk, Belarus in 1997.[10] The latter was, however, closed by Lukashenko's autocratic regime along with some 20 other private higher education institutions operating in Belarus during the 1990s. By 2001 only one private university remains, which will be discussed later in this book.

The forms that private higher education institutions take in particular countries differ greatly. In some countries, for example Hungary, the title "university" is legally protected. In these conditions the private higher education institutions operate as vocational higher education institutions, professional retraining institutions, or fall directly under the category of profit-oriented companies. In other countries, for example Belarus or Romania, attempts have been made, at least initially, to launch private multi-faculty research universities.

The Market of the Reforms
There is a price for East European market reforms, part of which is covered by local taxpayers. Another part can be expressed in terms of the real income losses that the majority have experienced, as well as in terms of shrinking social security. Many of the reforms have been supported by foreign funds. Popular neoliberal market

philosophy suggests that those paying for the reforms own the results. Many foreign agencies have been involved in attempts to save some parts of higher education in these economically difficult times or in facilitating some kind of reform. There are major actors, primarily the World Bank, who can afford to pay for system-level reform or, to be more precise, have the means to convince governments to implement certain system wide reforms: to reduce the number of staff, merge institutions, and so on, including the degree to which higher education is to be subjected to markets.

Reforms are also sold piecemeal. Actors who cannot afford to leverage system-level reform can purchase a disciplinary field, a part of that, or an institution, or a unit. Through bilateral agreements donor countries have supported various fields of studies, for example business administration. Germany and the Scandinavian countries have been involved in similar projects in the three Baltic States of the former Soviet Union, Turkey in Kazakhstan, Japan in Mongolia, and Iran's possible care for the development of education in Central Asia may worry politicians in some neighboring countries. Such support has its political and economic dimensions with variously more or less clearly identifiable interests. Still, given the current economic hardships, aid is highly appreciated. Each donor has its own priorities. Not all the fields of study attract the same attention. Business administration is certainly in a far better position to sell itself than, for example philosophy. One of the lateral consequences of the presence of multiple donors is the atomization of higher education systems and even institutions as each part follows its donor's traditions and its defined mission.

As a result of the piecemeal sale of reforms higher education institutions broke into legally highly independent units.[11,12] Subsequently the policy-making level is divorced from institutional realities as the Council of Europe report well-describes in the Estonian case. Institutions pursue their own interests while the Ministry of Education makes hopeless efforts to launch national higher education policy: "The Advisory Team found that, generally speaking, the members of the higher education community were not only rather uninformed about quality assurance procedures in other countries, but also about the planned system for their own country."[13]

The most interesting part of the higher education reform market is the market of private higher education institutions. According to Levy, private higher education is a logical piece of the overall reform package.[14] It would not be so logical if public higher education had expressed greater readiness or had been given opportunities to follow the needs of the changing labor market and renew its faculty. Public higher education has also been unable to meet the emerging needs of various groups: the needs of the newly rich for expensive and prestigious education; the needs of working professionals for retraining; and of those who simply want to have a diploma. Mistrust of the business community in traditional higher education, combined with diversified need for educational services has created an alternative sector of higher education.[15] In the latter one can establish a diploma mill with a few hundred dollars, have a multimillion dollar graduate school, or become the owner and president of an *international business school* for, say, 10,000 dollars.

In many instances, foreign donors are more interested in investing in new universities than those established under the former political regime, which are perceived as carrying the spirit of communist irresponsibility. Many believe that it is difficult, if not impossible, to change the traditional, large public institutions while the need for different studies in humanities, social sciences, as well as several professional fields seem to be evident. However, it is possible that many of the new universities were established not as a response to educational needs, but rather reflect the miscalculations of foreign donors or the irresponsible behavior of emerging local academic capitalists. Darvas admits that the availability of donor funding can encourage universities to initiate programs that are inconsistent with the institution's academic goals.[16] According to an earlier Council of Europe report, the emergence of new institutions is a lateral consequence of the attainment of other goals pursuing individual interests or implementing barely related state higher education policies.[17]

Donors may also pursue their own interests. Furthermore, if a powerful donor believes it has found a problem and begins to cure it the problem will be there: "If men define situations as real, they are real in their consequences."[18] Cash is instrumental in bringing phenomena from the world of Platonic ideas into reality. The problem with the universities that have been established as a response to available funding is that they may never start teaching, or their degrees are not recognized despite their assuming the titles of "American," "European," or "international" as many of them do. It would be difficult to say how many of the nontraditional universities have been established to meet the need for different kinds of studies and how many have been established by entrepreneurial academicians to attract short-term funding made available by donors or students looking for something more exotic than five years of dull study.

The wave of opening new universities is moving eastward. In November 1997, the then First Lady Hillary Rodham Clinton opened the American University of Kyrgyzstan in Bishkek, the capital of Kyrgyzstan, a former Soviet Union Republic in Central Asia.[19] The American University continues the operations of the American Faculty of the Kyrgyz National State University on a larger scale by providing international-level training in business administration and journalism in supposedly uncorrupted teaching environs.[20] In an environment well known for its flexible academic standards and despite its developmental difficulties, the new university is attractive for the best student candidates, who happen to come from the most affluent families. If it is able to support its title with accreditation in the country with the education it is claiming to represent it may become a significant player among the universities of Central Asia.

One may still ask whether making similar transplants can heal the sick bodies of national higher educational systems or whether they should be treated as tumors. A number of reports by the Council of Europe support the latter hypothesis. The current trend may also lead to substituting markets where system-level reforms, disciplinary fields, institutions, and units are sold to various agencies for system-level policies. Educational provision for students who cannot meet its full cost is becoming increasingly difficult when short-term financial gains outweigh educational goals. Uncritical submission to Western experts and nonexperts seems to be viewed as the easiest and most cost-efficient way of reaching world higher education standards. The negative side of this approach is that, contrary to prevailing expectations, the rest of the world is far from standardizing higher education to the level that the

Soviet Union once did. While the EU is trying to reach higher compatibility between the higher education systems of the member states through an indirect scheme, it will hopefully never apply the Soviet central planning model. As Levy showed, choosing appropriate examples from a wide range of different cultures, any possible approach can be argued to be leading toward the goal of world or international higher education standards.[21] Irresponsible manipulation of the term *international*, however, leads to the delegation of decision-making and national policy making to the agents least interested in the long-term consequences—randomly selected external agents and individual consultants.

The Market of Degrees and Qualifications

It has been recently argued that because of the financial, legislative, and administrative difficulties East European public higher education has been experiencing, traditional universities have been losing their teaching capacity through "brain drain" and inadequate resources.[22] The conditions required to reverse the external brain drain, improved salaries, and higher priority for research and education budgets[23] are far from being attained. One can of course argue that because of the activities the universities were involved in under the previous political regime—teaching scientific communism, Marxist–Leninist philosophy, history of the Communist/Socialist Parties, or taking care of the soundness of students' political views—the actual loss is not so great. The problem is not, however, so much related to the politically determined content of studies in the past than to the fact that universities, having a fraction of their previous resources, are under pressure to accept an even larger number of students. High unemployment and decreasing social security triggers a race for higher qualifications in the expectation that they will open a perspective for some kind of employment. As having a higher education qualification may in some instances be perceived to be an issue of existential importance, the pressure on traditional or nontraditional higher education institutions to award degrees is unusually high. In this respect the situation is qualitatively different from what it was a decade or so ago when higher education qualifications were more an issue of prestige and might have lead to a lower rather than higher income.

Following the emergence of different consumer groups, the market for products defined in terms of higher education qualifications is becoming increasingly specialized. For low-income, working-class families the most important issue is to have their children graduate from any university. Representatives of this group do not apply for academically prestigious or elitist programs in public or private universities. They may be looking for something practical and achievable—probably teacher training or traditional fields of engineering in the public sector, which, in the majority of countries is still free of charge. The purpose of study is to avoid outright unemployment now and possibly later. If a representative of this group cannot meet the academic standards of a public university, the alternative is to study at the lowest-ranking private university, paying not too high a tuition fee and receiving a non-recognized qualification. The lack of experience and incomplete information means that the true value of the qualification received from a low-caliber university may well only be understood when unsuccessfully applying for a job.

Excellent students from middle-class families are probably also competing for limited nonpaying study places in public universities choosing, however, the field of study more carefully.[24] Foreign languages, law, and economics could well be considered. In the event that a student does not qualify through the admission procedure for one of the limited number of state-supported students' places, most of the public universities nowadays offer a second chance—to pay a fee and still study. Whilst technically illegal according to constitutional law or education acts in the majority of countries this practice is widespread. Universities usually accept 20–30 percent of fee-paying students and use the generated revenues to increase faculty salaries. The latter fact gives the fee-paying students a particular value in the relatively modest university environment, directly or indirectly diminishing the significance of the studies of others.

Demand for elite education has been a powerful motivation behind the recently established private education sector. Many of the newly set up universities allege their status to be international, and although the statement is usually weak in legal terms and is often supported by little evidence, the prospect of belonging to a special community is appreciated among certain groups of students. A simple correlation exists between the charged fee and the status of the university among the newly rich families. This group of students seems to be looking for an environment that agrees with their social and economic status, and for a degree confirming it. The most common fields of studies offered by private elite universities are law, business studies, and foreign languages. Journalism also seems to have quite a high level of prestige.

The International Concordia University established in 1993 in Tallinn, Estonia lends itself as a good example of the latter group of universities. Attracting students primarily from affluent Estonian families, but also from those of the two other Baltic countries of the former Soviet Union, its initial foundation lay on a dubious agreement with a partner university in the United States of America and the services of visiting junior faculty members.[25] However, its starting position as the first blatantly elitist university and its rather ingenuous market orientation allowed it to gradually improve its financial standing and subsequently improve the standard of teaching in addition to establishing international contacts. The history of the International Concordia University, however, ended abruptly in Spring 2003 when it went bankrupt with approximately USD 2.5 million missing on its bank accounts.

The concept of the new elite may need some further clarification. It may well be the case that strong continuity exists between the previous *nomenklatura*, understood as a closed social group (usually high-ranking governmental officials and Communist Party members) having access to especial benefits and services under the former political regime, and the new elite that the new democracies have given birth to since 1989. The process of privatization of state property through which the former positions, memberships, and relationships have been translated into economic capital supports the above proposition well. If the elitist new universities are able to maintain their special status they will definitely have an important role in cementing the new social structure that may finally not be new at all.

The status of higher education degrees and qualifications is no less significant an issue for governments than it is for institutions and individuals. As Kivinen and Rinne put it: "In terms of education, upon closer examination 'the State' can be seen as a fountain of symbolic power, charged with overseeing the educational rituals associated with the award of educational certificates and diplomas."[26] The symbolic power of the East European states is fatally leaking away. While the market declares its right to control the certification rituals, the state is trying to maintain its role by trying to facilitate recognition for its loyal universities among the normal Western states.

Applying the metaphor of more or less successful introduction of convertible national currencies, the issue of introducing fully convertible degrees and qualifications ranks high in state reform programs. Ignoring the political and economic aspects of the issue, binding agreements with Western countries are sought for mutual recognition of degrees and qualifications. The immediate threat of a large number of unemployed, inadequately remunerated Eastern degree-holders entering the West European labor markets makes Western governments rather cautious. Still, efforts have been made to copy programs and structures from the EU countries and North America to strengthen the request for formal recognition. It is primarily for this reason that most of the countries in the region have shifted from the traditional long-cycle, five-year university program to bachelor, master, and doctoral studies, with mixed results so far.[27] However, it is no secret that politically driven top-down reforms have, in some places, for example the Russian Federation, been rejected not only by the local labor ministries that do not accept the new degrees issued by the public universities of their own countries, but also by some of the leading universities that refuse to issue them.[28]

Good marketing of degrees is vital for private universities whose revenues consist mainly of tuition fees. Unsurprisingly, many of them violate the principles of consumer protection in claiming their products have qualities that they actually do not. Many universities tell their students that they meet international standards, that their degrees have international accreditation or, if the leadership is more humble, that they follow European or American traditions. This usually either represents the leader's wishful thinking or it is a downright lie. A good proportion of private universities do not enjoy recognition in their host countries. Along with an inability or unwillingness to provide quality education, many of the recently established state-run accreditation agencies derive their concept of quality from the context of research universities. Applying it to short-cycle vocational schools or other nontraditional higher education institutions may not be fully appropriate.[29,30] Making accreditation impossible for recently established private higher education institutions may be one of the reasons why the procedure has been launched by traditional institutions and governments for new universities. As before, this protects the symbiosis of the state with its loyal universities.

Labor Markets

Eastern Europeans' love of markets and market economy is apparently a well-established fact. The extent to which market reforms are ends in themselves and how far they represent means toward other ends, for example access to foreign investments and funding, remains open to debate. Despite the market euphoria, a conflict can be perceived between the official discourse and the public perception of how the economy, as well as the public sector, should be run. The fate of politicians supporting the introduction of fee-based tuition in the public universities may well follow that of Socrates, who was condemned for charging his students a fee. Despite this some still share his view, which has been unsuccessful in Europe over the millennia: "The fact is that there is nothing in any of these charges, and if you have heard anyone say that I try to educate people and charge a fee, there is no truth in that either. I wish there were, because I think that it is a fine thing if a man is qualified to teach, as in the case of Gorgias of Leontini and Prodicus of Ceos and Hippias of Elis."[31]

People who have presented market reforms as almost a matter of life and death suddenly request that some of their fellow citizens—university professors—should be excluded from enjoying the fruits of the markets. With a few exceptions (e.g. Bulgaria and Hungary) legislators in the countries of Eastern Europe and the former Soviet Union have passed laws that stipulate free-of-charge higher education in public higher education institutions. In the context of growing numbers of students and of economies experiencing major difficulties, public budgets are able to meet low salaries and utility bills, but not always even that.[32] Depending on the country, the monthly salaries of faculty are around $400 in some Central European countries, $50–100 in the Eastern part of the continent, and about $10 in some Central Asian countries of the former Soviet Union. The fact that the public university faculty income, as a rule, does not reach average income levels in particular countries[33] largely defines faculty behavior in the labor markets: those who are able to find jobs anywhere else leave. People who have worked for universities for 20 or 30 years have, however, often no alternative. One of the few solutions to earning additional income is to lecture part-time at one or more of the private universities.

Privately established schools or universities are either not covered by the prohibition on charging students for their studies or operate as consultancy companies, foundations, and so on and ignore the law. As a result only a fraction of the private university faculty hold full-time positions. Even in relatively well-established private universities up to 90 percent of the faculty may be working part-time. This, on the one hand makes the private universities relatively cheap to run and keeps tuition fees at an affordable level; on the other hand, it makes it rather difficult to develop the institutions further. However, it may well be the case that higher education is moving globally in the same direction and that soon no university will be able to afford a large tenured faculty, or as Neave states, "in short, we must turn to higher education to make ourselves more competitive, our skills more saleable, for the simple and devastating reason that otherwise the rich can no longer afford us."[34] This is something that Eastern European faculty can teach the rest of the world and earn their living for another day. Being under double pressure, from the public not to introduce fee-based tuition and from the universities in demanding additional funding, most ministries of education are compromising the law by allowing the public universities to accept, and charge them more or less as they wish, around 20–30 percent of students in addition to those under the *numerus clausus* and covered by state funding. Year by year, the share of the fee-paying students is growing. In similar legal conditions, the University of Vilnius has, for example reached a level where 50 percent of those admitted in 1997 were fee-paying students.

The problem of sharing this additional income between the state, the university, and the teaching faculty is a difficult issue in conditions where, strictly speaking, the income cannot exist. The taxation office can always find a reason to get its share. Tensions have been created, however, within the university community itself on the grounds of sharing the income between those who can earn additional money, for example the Faculty of Law and the Faculty of Economics, and those who cannot, let us say, the Faculty of Philosophy. As one may anticipate, academic solidarity is not a strong argument in this dispute.

During the first years of the transition, additional teaching and income were usually made available by small groups of academic entrepreneurs who formally established private universities and contracted the teaching activities to their former public university colleagues. Recently, public university leaders have understood the destructive nature of the trend, as well as the hopelessness of the situation in overcoming communist ideals in higher education legislation. In some countries, most notably in the Russian Federation, public university rectors have started to take the initiative, establishing private parallel structures for their own public universities.[35] Despite its bizarre legal outlook, the provision allows control of faculty activities, as well as persuading the private structure to invest a part of its revenues on infrastructure development.

In the long run, East European public sectors are unable to sustain the large higher education systems that were once supposed to serve very different societies. Using limited legal means to prevent public universities from generating additional income whilst offering political support either accelerates the fatal process or corrupts the university community, or both. However, as the largest faction of the faculty is reaching the age of retirement and university positions are not attractive to the younger generation,[36] biology will help higher education out of the difficult situation created by economics. To paraphrase Stalin's infamous maxim one could say: When there is a university there is a problem, without a university there is no problem.

Market Educational Services

Conflicting reform initiatives and mutually excluding public policy decisions have led to extensive illegal operations and a further decline in faculty moral, which was never the strongest feature of communist universities, anyway. Having discussed some of the reasons for the growing demand for higher education qualifications and sources of new programs and institutions since 1989, the question about the relationship between these developments and teaching remains.

Bourdieu and colleagues have complained that university students understand little of what their professors teach them.[37] If this is correct, it might be one of the major factors that saved East European countries from everlasting communism: the students who later filled various leading positions in society did not understand what their professors had taught them about the inevitable victory of the World Communist Revolution, and used common sense in their professional life instead of what was taught as the deepest truth ever revealed to mankind. The fact that the same universities with the same faculty and even fewer resources are training leaders for very different societies confirms, however, that actual learning is no more important now than it was then. One may argue that teaching is never more than the tertiary function of a university.[38] The whole meaning of learning, beyond that of staying in a university for several years, can be questioned if the same education that was offered a decade ago is now provided with considerably fewer resources and less qualified faculty, and yet it is still considered good enough for supposedly dramatically changed societies. Post-communist societies seem to be trying the lowest level of allocation for higher education under which it still retains its credibility—not recognizing that the rise of formally nonrecognized universities indicates the closeness of that critical point.

When compared with the number of reform initiatives started because of the availability of donor funding, the demand for an elite environment, or the desperate need for additional income, the share of initiatives aimed at providing a different learning experience may not be overwhelming. One can find institutions providing retraining for degree-holders for recently emerged jobs in business and banking as well as in some other traditionally neglected professional fields like social work. A remarkable example within this group is the Moscow School of Social and Economic Sciences, which having found a legal niche in the Russian legal framework outside of the higher education system, has developed prestigious retraining programs in cooperation with the University of Manchester. The International Banking Institute in St. Petersburg offers short-term practical training to degree-holders who are looking for more prestigious or better-paid jobs in the recently emerged banking sector.

There is also a rather well-known group of so-called *Invisible Colleges* in various countries, which provide limited groups of outstanding students with tutorial training under the supervision of outstanding professors from particular countries, and additional learning resources as well as stipends. This initiative, recognizing the difficulties of boosting the quality of teaching on a larger scale and being critical about the academic intentions of the new universities, is still trying to support the development of the new generation of university faculty. It is definitely a very different study environment from what has been traditionally offered, bringing students and professors together for an extensive apprenticeship. However, as many of the programs operate primarily thanks to foreign donor funding it has a strong financial attraction for professors, as well as students. Despite having no formal academic standing within particular educational systems, the colleges promote a very clear elitist identity among the students and the chosen faculty. Unfortunately, it is not known how much attraction such programs would have without attractive student stipends and faculty honoraria.

Despite generally broader freedom, traditional universities have not been highly successful in diversifying their teaching activities. One of the first outcomes of the higher level of institutional autonomy was that universities abolished part-time and distance programs. The next step after the closure of the distance education programs was the emergence of private or semiprivate structures franchising foreign, for example the British Open University, distance education programs. Making no comment on the quality of the training provided through distribution of self-study packages, some people may relate this development to what happened a few years ago when fast food restaurants emerged as the heralds of civilization. Following

Apple, some people may prefer more substantial food than cheap French fries.[39] However, in times of change it is increasingly difficult to distinguish between the two. The actual price charged is not a good guide either.

Conclusion

Societies in transition have put their higher education in a rather difficult position when they demand: first, free-of-charge higher education to be made available for all who qualify in academic terms; second, demanding it to be provided at a recognized international level; third, expecting the result to be socially relevant for their particular countries. The only way higher education can respond to conflicting demands is to offer mutually exclusive solutions. This is not the best possible environment in which to reform higher education systems. One should add to the above the inherent interests of the faculty in maintaining their positions and social security under conditions in which unemployment may reach the 50 percent–level, when unemployment benefits are largely nonexistent and when lost jobs mean instant poverty. In trying to respond to market signals, higher education is falling apart with each of the pieces trying to meet some of society's conflicting expectations.

As a result, at the system level one can see a picture that is full of the symptoms of a particular mental disorder. In this picture, the state, having passed through considerable transformation has not been able to define its position as related to higher education. As this position is not defined, various governmental agencies follow their own best understanding. This is best exemplified by the example of nonrecognition of newly and officially established bachelor degrees in public universities by the labor ministries of a few countries. Certain elements of higher education are still regulated following the former rigid pattern, and new rigidities like state accreditation emerge, often with the support of external agencies. Finally Western, perceptibly omniscient experts enter and suggest the establishment of vocational higher education institutions as the OECD usually does,[40] merge the universities as the World Bank suggests, or give the universities lump-sum budgets as Council of Europe advises[41] without bothering too much that suggestions like these may not complement each other or with the local context. What cannot be decided upon because of conflicting interests, is left to the markets to regulate.

The rest is derived from the above background—public and private universities, often not recognizing each other's existence as institutions, use the same staff to the greatest extent. Politicians, as long as they represent their constituencies, vote for free-of-charge higher education whilst as parents they either send their offspring to often formally nonexistent elite private universities or pay for their studies at public universities, being not specifically concerned that charging a fee is illegal.

A variety of political impacts on higher education may be seen as important factors in stabilizing its excessive marketization. However, strong but unfocused administrative regulations, best exemplified by the Romanian example,[42] but common in the whole region, combined with an explicit demand for a much larger variety of educational services than was permitted under the totalitarian regimes is rather helpless in its attempts to force the spirit of the market back into its bottle. Even keeping it at the level of the Clarkian metaphor can only fail in the context of incomplete legal frameworks and significantly downgraded administrative structures. Recent experiences clearly demonstrate the effectiveness of administrative means in paralyzing reforms in public universities; however, there are more than enough loopholes in any of the current legal frameworks to enable the creation of alternative educational structures. The confrontation of the two types of universities may ultimately lead to a situation where public universities authorized by the state to award degrees lack the means and incentive to teach, and the private universities—both those truly devoted to teaching and the diploma mills—can only award nonrecognized degrees. Two aspects of higher education, teaching and certifying, seem to be hopelessly separated. In terms of the emerging highly specialized market, knowledge and certificates are defined as different products. To what extent the situation has been different in other times and other cultures remains to be known. Success in redefining the higher education sector and enabling the markets to play a well-balanced role in each of its parts may be decisive for the future of accessible and academically meaningful higher education in this part of the world.

Chapter 6
Reaching Beyond Geometry — The Privateness of Private Universities[1]

A communism which dissolves itself is a successful communism.
             Jean Baudrillard[2]

On the most general level one can distinguish between two groups of higher education commentators in Eastern Europe. First, there are those who declare that the necessary reforms have been successfully completed. From this perspective, East European countries have amended their higher education legislation, transformed the structures of higher education systems, and reformed study programs. The conclusion is that with all this being accomplished, the degrees awarded by East European higher education institutions should be fully recognized internationally for academic, as well as professional purposes, and that the countries themselves should be admitted into the EU. Second, there are those who say that there has never been much need for reform because East European higher education has always been excellent, perhaps the very best of all. The conclusion, however, remains the same as in the first instance.

In addition to these two groups there is a third, an insignificant critical group that tries to spread the view that although there is an urgent need to reform the content of East European higher education and change the academic culture, at the level of national higher education systems there has not yet been a success story from which to learn how to reform post state-socialist higher education. Lord Dahrendorf is perhaps one of the most respected people representing this critical camp and his recent book is quite clear on the extent to which higher education has been reformed since 1989. The list of Hannah Arendt Prize competition finalists include 18 institutions and initiatives. That several of them appear more than once suggests that the base of reform initiatives among which they could choose was not very broad. The 18 candidates may well represent almost the entire range of what one could possibly consider to be successful initiatives in human sciences in a region that covers the whole of Eastern Europe with the exception of the CIS. Summarizing his experiences of running the Hannah Arendt Prize program at the Institute of Human Sciences in Vienna, a project to identify outstanding reform initiatives in East Central European higher education, Dahrendorf concluded, "almost without exception, the institutions and initiatives which became Hannah Arendt finalists for their reform efforts are located at the periphery of the existing system of higher education."[3]

Reform initiatives in East European higher education are marginal. As Lord Dahrendorf also indicated, their impact on respective national higher education systems is insignificant. More often than not reform initiatives remain fragile. Quite often they depend on the initiative of a single individual. They also tend to depend on foreign funding. Being supported by foreign donors often develops a different scale of economy in the recipient program. Faculty remuneration is higher and institutional infrastructure is better. Donors' funds facilitate the building of libraries and computer networks very different from those that mainstream institutions can afford. All of which not only creates envy, it is often a source of political conflict. A recipient may be accused of selling national secrets to a donor who is accused of being, for example a CIA agent. Such incidents abound throughout the 1990s in many countries where the Soros Foundation operates.

In one way or another it is the invisible hand of the market that allows the establishment of innovations at the margins of national higher education systems. Such initiatives remain fragile and often contain all kind of conflicts caused by their marginal status, as well as by a hostile professional or political environment. This chapter first offers a theoretical explanation on introducing market mechanisms in East European higher education. After that, discussion is developed with reference to the case of the European Humanities University in Minsk, Belarus. This particular example can provoke some further thought, and particularly how far reforms in East European higher education are at all possible and what the motivation behind real or fake reform initiatives may be. As Jean Baudrillard notes: "Now, contrary to the apparent facts which suggest that all cultures are penetrable by the West—that is, corruptible by the universal, it is the West which is eminently penetrable. The other cultures (including those of Eastern Europe), even when they give the impression of selling themselves, of prostituting themselves to material goods or Western ideologies, in fact remain impenetrable behind the mask of prostitution. They can be wiped out physically or morally, but not penetrated."[4]

Applying Clark's Triangle in Eastern Europe

Social scientists, including higher education researchers, love simple models. Their love of circles, quadrangles, and arrows is particularly passionate. These seem to convey the hidden truth of things, unattainable to a simple, speculative mind. The purpose of designing models in social research is to create minimally complex tools with the capacity of generating the maximum range of future states. As with a formula in physics, for example one looks for the simplest possible tool with the maximum predictive power. This is, of course a difficult dilemma, particularly in social sciences. Too many variables or dimensions make a model difficult to understand if not meaningless. For example one may try to imagine a geometric model of six dimensions. By reducing the number of dimensions one may easily reach an elegant model that unfortunately does not explain anything. The latter has happened with some higher education researchers who have tried to apply Clark's famous "triangle"[5] to East European higher education reform. Clark's model states that higher education systems are coordinated from three possible points: state, market, and academic oligarchy. These three points make the three corners of his triangle. Any higher education system can be located within the triangle with the actual position being determined by the share of domination belonging to each of the three powers.

The non-reform of East European higher education has taken place under the guidance of a large number of foreign consultants. Since the 1989 revolutions, the European Commission alone has spent the equivalent of hundreds of millions of dollars on reforming the national higher education systems of Eastern Europe. It is no secret that a significant element of such funds is paid to Western consultants. Neither is it a secret that British universities whose funding has experienced significant cuts since 1979 have been encouraged, in particular, if not forced to go to the consultancy services marketplace and bring European money back home. And so the wise men from the West go out into the wilderness with their briefcases full of universal solutions and powerful explanations of how, for example to relate East European universities to industries that unfortunately may not be there, or to develop strategic plans while the only target for many of the universities is to survive another month, semester, or year. The consultants' experience, often drawn from a single country, creates a fertile ground for comparative curiosities not so far from Watson's example: "At times it is as if a comparative anatomist were to say that lions and ants are similar and are at a comparable level of development since both are warm-blooded, have six legs and are always winged."[6] It is, at times, only a short step from dogmatism to misinterpretation. To put it succinctly, Western consultancy is not always relevant. While this is by no means a secret, politics in the European Commission is a complex business.

In the view of the Western consultant, East European higher education, once highly dominated by state control has, in the 1990s, moved significantly toward the market. While previously the state controlled the institutions, it is now the market that dictates whom and how to teach and at what costs. There is some truth to it, meaning that in Eastern Europe there are institutions and institutional groups that operate precisely in this way. However, there are also a large number that do not.

From a methodological point of view, it would be difficult to apply Clark's triangle to state-socialist countries for the simple reason that under those conditions markets do not exist and considering the academic oligarchy—academics under surveillance by the security agencies—as a separate actor is highly problematic. Actually, it would be difficult to define any power-center for these higher education systems other than the total, omnipotent state. This means that, for a totalitarian country, the Clark triangle transforms into a circle as the state fully controls both the market and the academic oligarchy, either through detailed normative regulations or organizations like the Communist Party, security services, and many others.[7] Following this view, the most significant development in East European higher education is not the reorientation of higher education systems from the state to the market. It is the emergence of actors, a new kind of state exercising only limited power, the market, and the academic oligarchy, which can be considered as the corners of the Clark triangle. There is nothing wrong in saying that the communist higher education systems were state-oriented, but in contrast to many others, for these countries the state was the only possible orientation.

The first sign of emergence of other corners of the triangle is the formation of markets for higher education in the context of a continuously close relationship

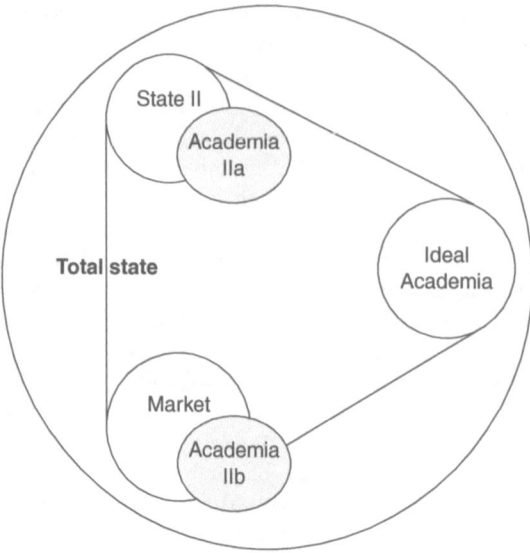

Figure 6.1 From a circle to a triangle.

between the state and the academic oligarchy. The latter, being unable to respond to the need for new professional fields of study and massive pressure on admissions, continue to play their own, old games. Furthermore, as a response to market signals, private higher education emerges. Alongside it, emerges market-oriented entrepreneurial academia (Academia IIb in figure 6.1). At an individual level many of the faculty belong simultaneously to the market as well as to the state-oriented groups.[8] Within this dual-identity, the relationship with the market generates the income, and the relationship with the state produces social status and basic social security. Many of the faculty experience a personality-split as they serve the state in the morning and the market in the afternoon. This is the reason for very peculiar circumstances concerning the recently introduced accreditation procedures where not only does everybody accredit each other, but also themselves.

As a result of the following three steps the totalitarian circle is transformed into the triangle of liberal market democracy:

1. Limiting the powers of the state;
2. The emergence of two independent centers of integration—the state and the market;
3. The creation of two loosely connected sectors within the national higher education systems related to the respective centers of integration and the faculty with their split identities.

This model would allow an explanation of some of the conflicts present in East European higher education, for example the conflict between the two subsectors both claiming the authority to certify the students, either on behalf of the market or the state. To handle the issue as state–market reorientation, leads to false expectations of a relatively easy transition. This is by no means the case. The system is clearly split. The great majority of the academic oligarchy is not ready to move toward the insecurity of the market that their governments may want to push it toward in times of economic stringency. East European professors, being traditionally married to the state, are quite unwilling to accept divorce without substantial compensation.

Trying to map higher education systems in Eastern Europe, one can see that for many countries, even in normative terms, two relatively separate sets of institutions exist. The first one consists of institutions established prior to the revolutions. These institutions have enjoyed a privileged position stemming from their relationship to the state and their mission of preparing caders of socialist/communist leaders. At one time, these institutions constituted a part of the former state apparatus. Although legal regulations might have changed that, many of the former practices continue. The second group, of independent institutions, has appeared only after political liberalization. Their legitimacy is derived from the market and is constantly challenged by both the state and academia from the state-dominated institutions. For various reasons, the market has not been able to fully legitimate the new higher education institutions to the extent that was expected. Both the institutions' own inability to offer high quality higher education, as well as the pressure applied by the public universities seem to play a role in this.

The conflict, however, does not lie so much on either the legitimacy of the groups in providing training or conducting research, as on the authority to produce symbolic capital, higher education degrees, and qualifications that can be legitimately exchanged for economic welfare and social prestige. As mentioned in chapter 5, in Eastern Europe certificate and education are not necessarily related. The problem that we are facing is related to the primary function of higher education establishments—to screen, select, and certify people for power and the highest security in particular national settings. To carry out this mission universities need a broad acceptance of the credentials they issued by their respective societies, from the governments allowing graduates to apply for jobs in governmental structures, from the employers allowing the graduates to apply for highly rewarded executive jobs, and so on. In his recent study Darvas describes traditional universities like the Charles University in Prague, ELTE in Budapest, Comenius University in Bratislava and others, as defending their monopolies.[9] The only detail missing, thus preventing Darvas from reaching the bottom of the current conflict, is that the protected monopoly is the monopoly of certifying the future elite on behalf of the state and to receive public funds to accomplish the task.

Until the late 1980s, higher education institutions exercised the function of screening and certification of the future elite on behalf of the communist regimes, applying nonacademic criteria implicitly or explicitly. Since then, the continuation of the same function by the same institutions, for presumably changed societies, has been challenged. During the early days of the Velvet Revolutions, radical solutions were proposed, including the closure of the former higher education institutions. Later, developments have shown this to be neither possible nor desirable. The changes that have occurred have been much more gradual than was initially expected and the *new* societies are considerably less new because of the vested interests and cultural problems that Baudrillard has mentioned—East European societies remain culturally impenetrable. The question of legitimacy has been institutionalized in the emergence of an alternative higher education sector that, in addition to questioning the monopoly of the traditional institutions to produce the credentials, claims a share of the public budget. For new higher education institutions the fact that the societies have fully recognized the symbols produced by the communist universities is a somewhat unexpected and unfortunate turn of the events. What we see happening is the polarization of higher education. One of its parts is still close to the state power while the other tries to serve capital alone.

Private Higher Education in Belarus

Attributing too much explanative power to abstract models may be one of the reasons why our understanding of post state-socialist higher education lacks the necessary depth. In his apparent rebellion against the media guru Marshal McLuhan, Baudrillard diagnoses the situation as *la confusion du medium et du message* [the confusion of the medium and the message].[10] The limited number of possible moves within a triangle may not be sufficient to express adequately the underlying complex reality. The structure of the model largely defines the content of the message to be delivered. One of the few things worse than taking a model too seriously is not taking it seriously enough. When this occurs, insights such as the following are the product: "Referring to...classification presented as a continuous scale ranging between the extremes of a collegiate model towards an entrepreneurial system..., the German University might be located within the middle of a 'magic' triangle formed by three corners: scientific oligarchy, state administration and bureaucracy, and market orientation."[11] It is not fully clear whether Clark ever considered a non-Euclidean interpretation of his triangle such as the one presented here in which the author has mixed two different models. In the first part of the sentence he has perhaps unwittingly modified another of the favorite models that higher education consultants use, while in the second part he is looking for the third corner of a line.

Clark's model has its obvious limitations, as any simple model in social sciences inevitably does. Instead of discussing it any further, the issue of private higher education in Eastern Europe will be further discussed later using the example of the European Humanities University (EHU) in Minsk, Belarus. The EHU is one of the many *Western* transplants in East European higher education, which came about largely as the result of generous foreign donations. The theoretical framework for the following discussion is borrowed from Daniel Levy's work on Latin American higher education.[12] Reading what follows, one may wish to do what Timothy Leary tells us many have done before, namely "diagnose entire societies as paranoid."[13] That, however, is not the purpose of this chapter. Rather, it is a modest attempt to reveal a small part of the post-communist reality whilst trying at the same time to avoid serving anybody's immediate political interests.

Like developments in other former Soviet Union countries, Belarus experienced significant private initiative in the provision of higher education immediately after the breakdown of the Soviet Union in 1991. In the years 1992–96, in total 20 *non-state* higher education institutions were licensed by the Ministry of Education and

Science.[14] The development over the last few years has been toward limiting the number of independent higher education providers through the process of state accreditation under the neo-totalitarian political regime. The latter process narrows down the differences between state and non-state higher education through compulsory unification of academic programs and institutional structures.

Of the 20 non-state higher education institutions established between 1992 and 1996, 14 were still operating in 1998. Of the six discontinued institutions, one did not admit students after licensing, two were closed shortly after opening, two were taken over by state institutions in whose premises they operated, and one was closed as a result of unsuccessful state accreditation. Out of the 14 institutions that remained, nine have received state accreditation for at least one of their academic programs. The five remaining institutions operate under the threat of immediate closure and possible legal action by students and their families. Of the nine partially accredited higher education institutions only one, the EHU has been granted the prestigious status of a university. The rest are *institutes of higher education*, referring to their lower prestige and narrower academic profile. It may appear a significant achievement for the EHU to exist as the only university-type, non-state higher education institution in the Republic of Belarus. But, the status has its trade-offs. Some of them may actually question the very idea of establishing a university like this—creating an environment for autonomous teaching and research detached from the political past of the country and an ethically corrupt state higher education environment, seems to be an almost impossible task.

The Privateness *of the European Humanities University*

The distinction between public and private sectors in higher education is not as clear-cut as one might expect. Higher education legislation does not always distinguish between private and public institutions, and even if it does, the formal legal definition does not always fully explain the mechanisms by which particular institutions are controlled and managed. In cultures where a strong activist state has for decades been fully responsible for running the higher education systems, minor steps toward limited *destatización*[15] could easily be misinterpreted as privatization of the whole sector. Alternatively, far-reaching market reforms in commerce and industry may encourage the privatization of higher education institutions rather than less radical reforms that would pay due respect to the public functions of higher education. The conceptual framework for thinking in this way in post-communist Eastern Europe is not yet well developed and therefore extreme solutions tend to prevail.

From the funding point of view, a university that is clearly identifiable as private remains a rare phenomenon. Internationally, many private higher education institutions rely to a significant extent on public funding. The converse is also often true, private source funding in public universities may also be highly significant. The latter is true in Eastern Europe, funding of private universities here is usually more private than in most other countries, which may well explain the low profile of many of these institutions. As described in chapter 3, some countries, notably Estonia, have tried to reach *destatización* through the redefinition of the actors in the legal system, which, at least in principle, limits the gap between the public and private higher education institutions.

In the majority of the former Soviet Union countries, including Belarus, the legal status of so-called *private higher education institutions* remains unclear. While being called *private* colloquially, the legislation in these countries does not usually refer to private higher education. The legal status of the EHU in Russian is *ne-gosudarvstvennyi'* meaning "non-state" reveal actually what the university is not rather than what it is, that is, it is *not* a state university. To understand the true meaning of the *non-state* status and its implications on the operations of an institution, further analysis is needed.

The founding documents of the university are not very helpful in this matter, either. The EHU Founding Agreement provides a rather diverse list of individuals and entities that have joined to establish the new university.[16] All together there are 13 founders. The list includes the Ministry of Education, the Ministry of Culture, the Orthodox Church, the Academy of Science, other governmental and nongovernmental organizations, as well as individuals, among whom is the current rector and one of the vice-rectors of the university. The list, including individuals as well as governmental and nongovernmental agencies, looks like a well-calculated political shield, particularly if one considers the internal political conditions of the country.

By avoiding any further development of conspiracy theories, one can at this point conclude that the simple public–private dichotomy may not be of much help in understanding the nature of new universities in Eastern Europe. As Levy states, "Abandoning dichotomies, it becomes painfully clear that private institutions are not even necessarily more private than public ones by matter of degree."[17] To further analyze the balance between privateness/publicness of the EHU three aspects of that dichotomy introduced by Levy will be applied: (i) finance, (ii) governance, and (iii) function. After that, a short analysis will follow on the extent to which these three criteria have an impact on freedom, choice, equity, and effectiveness as experienced in the case of the university.

On the nature of financing a higher education institution Levy concludes, "an institution is private to the extent that it receives its income from the sources other than the State and public to the extent that it relies on the State."[18] The financial status of the EHU is not common in the world of higher education—it receives not a penny from the public budget. On the other hand its budget is tiny compared to any mainstream university. The only way to run a university with some 700 students on an annual budget of one million dollars is to rely largely on the services of part-time faculty, a common characteristic of post-communist private universities. Currently, the 700 students pay annual tuition fees of an average 1,100 dollars, which is not a small amount of money in a country that has virtually no economy. Donor funds are, however, available to cover the tuition expenses of a certain number of students, as well as the general running costs of the university. In parallel with the establishing of the EHU in 1992, a private foundation, the International Humanities Foundation was set up. Although its charter states that the foundation's mission is broader than raising funds for the university[19] it is apparently not far from it. It is an interesting coincidence that the director of the Foundation is the very same person who is serving his second term as the rector of the university.

Difficulties in public sector funding force state universities too to look for additional revenue sources. While public higher education in Belarus is legally free of charge, state universities have been gradually introducing direct or indirect student fees. The Ministry of Science Education has tolerated the state universities breaking the existing law and admitting 20 percent of students—those who do not rank sufficiently high in the entrance competition, but who are willing to pay for their studies through fees. There the importance of private source funding in university budgets is growing and universities, in the dimension of their funding, are becoming increasingly private. Even in this context the position of the EHU is rather extreme since on the basis of its funding scheme it is a 100 percent private university.

On governance, Levy's view is, "An institution is private to the extent that it is governed by non-state personnel and public to the extent that it is governed by the State."[20] The EHU's governing structure combines public and private features. The overall structure of the university seems to replicate that of state universities. The higher governing body is the University Council chaired by the rector. The rector also chairs the Academic Council. In this way the rector is the academic as well as administrative leader of the university. There is no separate administrative structure. At the faculty level, deans carry out dual functions as well.

The issue of governance becomes more interesting if one searches for the source of the rector's power. In state universities the Rector, formally elected by the University Council, is appointed by the Ministry of Education. In the case of the EHU the Council of Founders somehow occupies the role of external supervisory authority, albeit rather ill-defined. The Founding Agreement stipulates that the functions of the Council of Founders are defined by the University Charter.[21] However, the only thing the University Charter actually says about the role of the Council of Founders is that a representative authorized by the Council contracts the rector elected by the University Council.[22]

Currently, the individual who serves as the first rector of the EHU is also among the founders and at the same time serves as chairman with the casting vote in the University Council. One can easily imagine a situation where a person on the one hand elects himself as the rector and on the other, as the representative of the Council of Founders and signs a contract with himself. When one considers that the current vice-rector is also a member of the Council of Founders even less may suffice—mutual agreement could easily substitute for self-contracting. The way in which the EHU governing structure has been established does not necessarily make it public. Given the overlap between members of the Council of Founders and the University Council, it effectively allows the isolation of governance from any external influence, making it private and open to an excessive concentration of power in the hands of the key founding members of the university. This is by no means unique to *innovative* higher education institutions in Eastern Europe and perhaps beyond. It is almost the rule that these institutions have been established by reform-minded and charismatic individuals. In a sense, their way of pursuing reform cannot be termed democratic. The real crisis strikes when the leader changes for one reason or another. Although perhaps Dahrendorf does not see the whole problem of pursuing democratic reforms by undemocratic means, tending to romanticize the personality of such reformers, his experience with the Hannah Arendt Prize nominees and winners clearly indicates that such initiatives are often fully dependent on a single individual: "... someone often stands out, as an inspiration, an engine of action, a leader, often a fundraiser and a propagandist for the institution. Moments of reform are moments of personal leadership .... when the founder leaves the foundation is at risk. More often than not it will either close down or fall into a routine..."[23] Finally, one cannot decide on the nature of a university without considering its core activities. Here Levy defines his use of the term function, "by function I mean, most simply, what the University does. Less simply, function refers to the ramifications of what is done, especially what interests and values are served."[24]

Whilst Belarus faces extreme economic difficulties and the country has to accept private funding in higher education, it is not too concerned about governance as long as the formally required bodies are in place. It has developed sophisticated procedures to keep core activities under strict control. Belarus has adopted external quality assurance procedures closely following those developed in the Russian Federation. These consist of three basic steps—licensing, attestation (examination of the *quality* of students and graduates), and accreditation. As in the Russian Federation, in Belarus accreditation is also necessary for a higher education institution to award state-recognized degrees and qualifications.

Training at higher education institutions, in public as well as in private, follows the same principles laid down in the State Higher Education Standards. This set of documents lists all the study programs that higher education institutions can possibly offer. It also lists all the subjects taught, issues to be covered, and number of hours allocated to each of the subjects. To meet the accreditation requirements institutions cannot omit anything the Standard prescribes. Additional subjects can, however, be added. This often proves to be difficult as teaching the Standard program already involves more than 30 contact hours a week. Looking at the State Higher Education Standards as they have been compiled in Belarus or Russia, one might easily conclude that officials from the Ministries of Education overestimate their own academic competence in drafting such documents. This is, however, usually not the case as the author of these documents is not usually the Ministry of Education or a similar body. It commissions the documents from groups they consider to be the leaders in particular fields of study in their countries. Therefore, the EHU does not teach Marxist political economy as a part of its economics program because the Ministry of Education and Science is particularly keen on it, but because this is the way academics from the Belarus State University have written the particular Standard.

In addition to measuring compliance with the strictly prescribed teaching process, the Ministry of Education and Science has developed a list of 25 criteria for institutional accreditation against which the whole institution is evaluated. Appendix 1 provides the translation of this most interesting document. The very first item in the list is most informative about the basic approach the document has adopted, according to which *international recognition* of an institution as *a center of higher education and research* is a necessary, however, not yet sufficient (there are 24 other items) requirement for recognition as a higher education institution in the Republic of Belarus. The only way to comply with this legally meaningless requirement is to endeavor to obtain a confirming document from the same organization that prepared the list of the requirements—the Ministry of Education and Science. This is precisely the manner in which the EHU solved this and other similar problems. By and large, compliance with the requirements is not assessed, but negotiated.

It would be fair to say that the EHU has tried its best to make the orthodox programs more acceptable for the students to study, as well as for the faculty to teach. However, as this can be achieved only by adding additional courses to already heavy study-loads, the immediate outcome is a high dropout rate among students who are often expected to spend 50 or more hours a week in the classroom under direct instruction. Despite all efforts, the functions of the EHU remain largely public—as for core activities, deviations from the standard teaching process and content are not tolerated.

To conclude the discussion about identity, the EHU resembles St. Paul who was a Jew for the Jews and gentile for the gentiles or in this case, carrying public functions to satisfy the accreditation criteria established by the Belarus State, maintaining a private governance structure to please nongovernmental and foreign stakeholders, and private finances for the simple reason that the state is not interested, at this point, in accepting any further burden on its budget. The price paid for private finance and private governance is that it functions as a public university. From the institutional point of view this is nonsense. It makes no sense to spend private money in order to carry on functions that the state would carry out anyway. This is true for external funding agencies, as well as for students paying impressive tuition fees. Despite the university's functioning very much as a public institution, it also has to carry out some marginal functions that make it attractive to private funders. There are perhaps political reasons to support an alternative university in one of the most authoritarian of the post-communist countries in Europe. The student body largely represents those who, because of heavy competition for free-of-charge higher education in the state sector, have had to choose the fee-paying track. For the faculty, simply in order to continue teaching, at the Belarus State University for example they need second or third jobs, preferably academic.

The implications of various dimensions of privateness on four key concepts: freedom, choice, equity, and effectiveness[25] of the EHU will be discussed next along the lines developed by Levy.

The Ramifications of the EHU's Status

Freedom covers institutional autonomy and academic freedom. This may sound anecdotal, but it is apparently a fact that traditional state-run higher education institutions in Belarus enjoy a higher level of institutional autonomy than private institutions. The state's active stand against non-state higher education institutions has made their position concerning institutional autonomy painfully clear. While the remaining private institutions, being under the threat of sharing the fate of those that have been already closed, can do naught but carefully follow the detailed rules, particularly those determining the content of studies, while public universities continue to enjoy their long established relationship with the political regime even to the point that violating existing laws, is not considered to be a problem. One high official in the Ministry of Education and Science accepted, during a conversation with the author, that practices such as charging fees for students in public universities is illegal, but is justified by the realities of life. During another conversation the vice-rector of the largest public university admitted that universities have, everywhere and always, stood close to political power and that he saw no problem in his university's special relationship with the regime lead by Mr. Aleksandr Lukashenko.

Earlier in this chapter it was argued that public higher education in Eastern Europe has established a strong alliance with the state that protects it against the market. The private sector is trying to draw its legitimacy from the market by directly satisfying market demand for higher education, and considers state regulation and intervention in higher education matters inappropriate. Fair and equal competition between higher education institutions is hardly possible when the state takes over the market, as was the case in the state-socialist countries. As a sign of continuity, in Belarus the government is also playing a large regulative role. Consequently, the EHU finds itself in an unfavorable position—the state, being paranoid about any bottom-up initiative, would rather see it disappear despite its largely public functions; markets have been systematically regulated to the extent that once again government officials produce *market signals*. The freedom the EHU still enjoys is neither derived from the state nor from the market, but from the not-yet-regulated gray area where the university leadership can pursue its own interests—academic or other. The university is working out its survival by playing the market against the state and the state against the market, perhaps testing the limits of tolerance of both. The freedom to do so may, however, evaporate as soon as the loopholes in legislation and other normative acts are closed.

One fundamental issue a private university in Belarus faces is that there is no meaning in what it is doing when it offers courses according to the state Standard. However, this is one of the basic requirements of receiving state accreditation without which the institution will be discontinued. As a result of the recently introduced state accreditation process, many higher education institutions have been already closed. The price paid for continuation is a difficult compromise. First, students have to take courses predetermined by the state. Providing the necessary minimum required by the State Standard for Higher Education requires approximately 30 weekly contact hours. In addition to that students are expected to take courses defined by the university, those actually carrying the mission of the university. As a result, EHU students are obliged to attend on average 47.2 hours of lectures and seminars weekly.[26] The EHU leaders strongly believe that all its graduates should be fluent in two foreign languages to be chosen from English, German, and French. Because of the very low level of foreign language teaching in secondary schools students take 24 hours of language training a week during their first two years of study at the university. During these two years 40 percent of the students drop out.[27] Students also study many new *Western* subjects, for example gender studies, in addition to the soviet-style programs without which the university could not exist. Examining students' knowledge on the Standard subjects is a part of the accreditation process. It is only after studying Marxist political economy that an economics students can have some micro- and macroeconomics.

The freedom students have to pursue their own ideas at the EHU does not seem to be any greater than at traditional state universities. While the EHU does offer courses not provided by other universities, for students it all comes as a predefined package designed by the university; students have no freedom to elect courses and they either follow the predetermined schedule or drop out. Ultimately, the dropout rate exceeds 50 percent. One of the main reasons for this is a conflict between the university's formal mission and its position in the marketplace. The EHU pretends to be a clearly elitist institution offering academically demanding (one may say that unnecessarily demanding) and expensive programs to the best students. In reality it is a not fully recognized institution, one where many of the students who cannot get a free position at the Belarus State University find their last resort. These students, however, are neither able nor motivated to follow the requirements applied in the university.

In relation to choice, the question to be asked is about the diversity an institution like the EHU brings to the higher education system. Do prospective students have more choice with the EHU than without it? As far as the Belarus State is concerned the question is irrelevant. Its position is that there is no need for diversity. All higher education institutions should be equal, providing standard courses and offering the same employment opportunities for all graduates. To a university that envisions offering an attractive high-status alternative to a unique group of students, high achievers from high socioeconomic status (SES) families, it may appear that it can make a difference, injecting more choice into Belarus higher education. The question is how far is it possible in practice given the conditions. The recruitment base as defined seems too narrow to sustain the university without public subsidies. There also appears to be a gap between the status declared by the university and the status perceived by the public. The EHU definitely brings more choice to Belarus higher education. But, the nature of that choice may be different from that which is claimed. From the students' point of view, the EHU offers an alternative. There are those, apparently a minority, who choose the EHU because of its richer academic program. For many students, the choice to be made in the midst of economic turmoil is between doing nothing (having not been admitted to public universities) or trying EHU for a while.

Turning to equity, the EHU does not add much to the existing system. Attracting high SES students, hardly offers additional opportunities to the excluded. The EHU is competing for the same group of students as the most prestigious state universities. Its lower prestige and economic status leave it with those students who can barely make it economically or academically. For the university, its highly demanding academic program is the highest priority although it is difficult to maintain under the existing pressures.

Considering all the above-mentioned factors obviously the EHU cannot be an effective institution. Despite its relatively low cost, a direct consequence of the fact that 90 percent of its faculty members work part-time, the educational process is wasteful. Students to whom studying at a highly demanding university is not a result of free choice, but the last resort before unemployment may be wasting their own resources in attempting the impossible. For the same reason the university is in a very difficult position in carrying on its primary mission, providing academically high-quality education in humanities and social sciences.

It may be that for the time being, the EHU is still sufficiently effective for its founding fathers—two philosophers who left the Institute of Philosophy of the Academy of Science to start their own university in 1992. Both of them are members of the Council of Founders, and the University Council, one serving as the rector and the other as vice-rector of the university. Probably for them the EHU still offers a reasonable alternative, particularly in comparison to dealing with philosophy within the official Soviet system. The margin of relative freedom one has as an intellectual in Belarus is, however, being reduced close to the level of Soviet Union. It is also true that EHU leaders have been able to use their positions to pursue their own academic interests in organizing and attending conferences, as well as running a publishing program translating Western classics. Needless to say, the EHU also serves as an additional source of income to a good many of the best faculty in the Belarus State University and researchers of the Academy of Science. How far this justifies running the conflicting and controversial teaching program is perhaps a matter of personal opinion. Ultimately, as was the case in state-socialist countries, there exists a level of political pressure beyond which independent intellectual inquiry becomes extremely difficult to pursue.

Conclusions

There are limits to which simple powerful models like Clark's triangle can explain the post state-socialist transition in East European higher education systems. Models can be modified by introducing additional variables and dimensions. However, the problem is that very soon they may lose their elegance and shortly after, their intelligibility. The only way out of this situation is to apply various models simultaneously, admitting that all of them are always only simple abstractions from complex realities.

Despite the illusory triumph of the market, higher education remains one of the most controlled sectors in the entire East European region. Belarus offers one of the most extreme cases. States that have recovered from their crisis in the days of anti-communist revolutions are turning their main weakness, a lack of funding, into their strength and making markets pay for their populist policies. This chapter has offered a somewhat extreme example from one of the politically and economically challenged countries in the region, using the framework Levy developed in studying higher education in Latin America. Unfortunately, not much comparative research has been done between East European and Latin American higher education systems. This could, however, prove to be a rather rich and productive field of study.

The case of the EHU provides a colorful example of how the state turns the aspirations of alternative academicians and their foreign donor support to their own benefit. Through its administrative machinery the state is powerful and can easily control the systems through *new progressive, Western-style* procedures such as accreditation, often promoted quite uncritically by international agencies and Western partners. Nonrecognition of degrees awarded by rebellious universities is a cheap but extremely powerful policy tool. The result is that, as Levy found in Latin America, private institutions can in some aspects be more public than the public ones.

Conceptually, the role of the faculty, in this case part-time, EHU faculty, remains a puzzle. Recruited from both public and private subsystems they simultaneously develop state standards providing higher education with a straitjacket, and for the rest of the day, while working at a private university, they try desperately to break free from it. It should cause some kind of cognitive dissonance, but apparently it does not. Maybe this somehow relates to the dual morality developed under the communist system. To be fair, one must add that capitalism's record in spreading sanity is not much better. Cognitive dissonance might be a normal condition for any human being struggling for survival.

Finally, is there a real division between the three corners of Clark's triangle? Before turning on, tuning in, and dropping out, Timothy Leary answered that question:

... we do not imply that there is "really" or "eternally" such a structural division.[28]

Appendix 1

Criteria for the determination of the status of university-type higher education institution, adopted by the decree of the Minister of Education and Science of the Republic of Belarus, N 26, 23/01/1995.

1. International recognition of the educational institution as a scientific (creative) scientific–methodological center of a particular profile.
2. Existence of scientific-methodological schools.
3. Existence of research training programs and examination boards for defending Candidate of Science and Doctor of Science theses.
4. Students' participation in research work with the aim of developing independent productive thinking. Operating systems of students' research activities, awards won in international or national students' competitions.
5. Share of fundamental research including national scientific–technical programs of no less than 30 percent of the total of research activities.
6. Inclusion in the content of training general scientific and general professional disciplines necessary for the training of Bachelor and Master degree-holders, and also specialized courses that draw developmental perspectives of particular fields of knowledge.
7. Total number of students no less than 3,500.
8. Of the faculty positions 60 percent should be filled with holders of academic degrees and academic qualifications.
9. For every 150 students one Doctor of Science degree-holder, recruited as a full-time faculty member.
10. Existence of a multidisciplinary profile of training programs similar to classical universities, and coverage of no less than 80 percent of the programs provided in the list of programs for particular types of profiled universities and academies.
11. Annual redesign of special courses of programs with the aim of including the latest achievements of science and practice.
12. Use in the training and retraining activities of progressive methods of training.
13. Offering in-service and retraining programs for experts and managers, according to the profile of the educational institution.
14. Effective training of experts with the highest qualification through the *aspirantura*[29] and the doctoral[30] programs (no less than 60 percent degrees awarded within three years after the completion of studies).
15. Acting as a leading contractor of strategic and fundamental research programs of the republic.
16. Presence of the fields of fundamental research where the higher education institution is the leader and recognized by the world research community.

17. Implementation of the main state-contracted projects of the development of science and technology.
18. With an aim of continuous maintenance of high-qualification involvement of all teaching staff in research, creative, or methodological activities.
19. Existence of modern research infrastructure enabling the preparation of experts with the highest qualifications, including Candidates and Doctors of Science.
20. Library resources with no less than one book per three students in every subject taught.
21. Presence of databases, modern scientific and educational literature, and periodicals covering the fields taught.
22. Compilation of new textbooks and teaching aids for the educational system of the Republic.
23. Presence of copiers and printing capacity for rapid publication of scientific works, textbooks, and teaching aids, and copying of scientific and educational materials.
24. Meeting the current requirements for lecture halls, laboratory space, dormitories, recreational facilities, and facilities for cultural and social activities.
25. Annual renewal of educational and research equipment and infrastructure in the extent of at least 5 percent of its balance-sheet value.

Remark: Among the criteria 7–25 the higher education institution reserves the right to choose 15 criteria.

Chapter 7
Exploring the Limits of Entrepreneurial Response[1]

The Parliament of Iceland, the Roman Catholic Church, and the university are the three oldest continuously operating European institutions. Here little can be said about the historic importance of the first. Considering the latter two, however; the inclusion of Latin quotations into the inauguration speeches of post-communist university rectors certainly strengthens this otherwise shaky hypothesis. At one point in the early 1990s, an Estonian university, through a simple misunderstanding of certain historical events, related its charter to the 1289 (!) Bologna Magna Charta of European Universities (the actual signing took place in Bologna in 1988). It was attempting to boost its own legitimacy, quite apart from restoring the identity of the medieval institution of learning by turning the clock back 700 years.

To this very day, the Roman Catholic Church and the university bear some significant resemblance that derives from common roots. However, there are also significant differences. While the former continues to recite: *"In sudore vultus tui vesceris pane, donec revertaris in terram, de qua sumptus es"* [In the sweat of thy brow shallt thou eat bread, till thou returnest unto the ground; for dust thou art, and unto dust shallt thou return][2], the university has its own sacred texts too. Administrators from international and intergovernmental organizations and the ever-growing ranks of higher education management consultants are in a powerful position to canonize texts or excerpts then to be read at international conferences, seminars, and training sessions much like prayers. Recently the American sociologist Burton Clark once again attracted the attention of European higher education management activists who realized that certain parts of his latest book fit well with their governments' agenda of pushing higher education into the marketplace. And so, quotes such as the following are recited from Salzburg to Sarajevo. Relevance, however, is not always the first priority of discussion: "Located in the northern periphery of European higher education, the modest University of Joensuu in rural Finland has come a long way in short time. A new university of the 1960s, it took up a special role in the late 1980s that led to a strong assertion of distinctiveness."[3] The principal difference between this quote and the one preceding it is that it does not conclude what the ultimate meaning of it all is. Transformation into entrepreneurial bodies clearly signifies the end of an important period in the history of the modern European University and its public intellectual mission. The university can continue to earn its daily crust with sweat and tears, but it certainly cannot preserve its purpose.

The current discourse of entrepreneurialism in European higher education has an inherently positive and mobilizing character. It is reminiscent of tales of the colonial conquests of the Wild West and of colonists taking care of themselves in a hostile environment. The message of the *entrepreneurial university* is also about taking care of itself in hostile conditions, conditions where governments lack the funds to secure the well-being of the university. The lesson universities can learn is not dissimilar to the one an infamous Soviet biologist, Mitchurin, preached some 80 years ago—one should not to wait for alms from nature, but go and claim one's fair share.

The Entrepreneurial University

Professor Clark's latest work rapidly became more than a study on changing higher education institutions. Falling like rain on the desert of European higher education, it has become a scientific-sounding justification and mobilizing force for turning it into the fertile ground of the marketplace—work enthusiastically welcomed by politicians. While administrators and bureaucrats ardently paddle the politicians' boat, academics hesitate. Instead of working under the guidance of the philosophy faculty and the protection of the state,[4] they are now forced to produce *knowledge* for sale.

Neave once wrote that this is the process by which the American dream became the European nightmare.[5] The idea of the entrepreneurial university is being sold under the pretext of progress and its pioneers heralded as colonizers conquering the Wild West. Agreed academia has to be mobilized to accept change on such a scale. But the related campaign looks very like mass-manipulation through *science* and some middle-level values in the Maslow hierarchy, and at a time when, for the bulk of mass higher education institutions, no alternative exists. Yet, so far one task that higher education research has not handled well is to lay bare the forces that have led to a situation whereby remaining non-entrepreneurial is a non-option. Apart from the anticipated digital utopia other arguments remain relatively weak. It may be difficult to accept, but the way learning institutions have been socially constructed and then massified under public pressure for social mobility leaves little room for learning as a student's primary motivation or for teaching as a professor's goal in life. Each strives for something else. Although both would prefer the state to come and pay up front, one cannot be sure whether this is exactly what is meant by social justice. Instead of being a source of public good, higher education is increasingly viewed as an item of private consumption with all the consequences that follow therefrom.

Still many believe that in the university students study while professors teach them. Based on this naïve, albeit sincere belief students in universities obtain knowledge and skills, which they later use for the benefit of themselves and their respective societies. However, what matters in terms of students' careers is not what knowledge they might posses so much as the symbols that represent it. The two things are not necessarily related. The behavior of students in universities might vary greatly depending on whether they went there to learn, earn a degree, or just to have fun. Pursuing the latter goal, for example would not even develop survival skills. Gellner asserts that the situation in modern, industrial society is still not beyond hope: "This modern growth-orientation has one immediate consequence: pervasive social mobility...Apart from its instability, in some measure at least, to be meritocratic it must fill some posts at least in terms of the talents and qualifications of available candidates."[6] At least some of the positions should be filled with people able to perform their tasks, which places certain limits on the automatic reproduction of the nobility in elite schools as described by Bourdieu.[7] In many cases social mobility is not so much perceived as a by-product, but rather as the primary goal of attending university. When they attend or send their children to universities many students and their families seek social mobility and the symbols that facilitate it. However, if the OECD ideal that Skillbeck, expressed as higher education for everybody,[8] is achieved, social mobility will be either facilitated by the relative status of degrees and universities or new means will be invented. If not, in the worst-case scenario, the whole process will be controlled by social capital, that is by connections and ultimately by wealth.

The games academics play in their committees and commissions have been much discussed. An important issue to consider is that the teaching they do—or do not do—has only a limited impact on the performance of students. Often it is teaching assistants who do the actual teaching while professors read their papers to each other at international conferences.[9] Academic tribes are exclusive. Each possesses its own inaccessible professional jargon and *high culture*. In this respect, the much-praised peer-review may well be seen as a means of group protection against public scrutiny. The *old boys* publish each other's papers in each other's journals and books. Colbeck demonstrates that in the case of higher education research, newcomers in the field are forced to adopt the same stance, often abstract and divorced from practical issues, studying and writing their papers in the same jargon to receive their tenure or risk never having a university career.[10]

One could conclude that universities not only face hostile demands for public accountability, but that public concern may be well justified. Much is going on in higher education that the public does not need to pay for, the bill going instead to those who play those games, either for fun or private benefit, because as Neave and Van Vught think perhaps rightly, society simply cannot afford it.[11] For society, the major policy issue is whether to choose between supporting students or supporting the growing number of unemployed. Cost is arguably crucial here, which may sound cynical, but in a society with universal higher education, unemployment benefit may to large extent determine the amount society can afford to pay for a university student. The margin between this and the real cost should be the responsibility and risk of students themselves.

As Clark says, "becoming entrepreneurial," or as Davies termed it earlier "adaptive and entrepreneurial,"[12] grants a university significant freedom from state control. However, the university only operates as the guardian of national high culture as long as the nation pays for it. This is a serious issue for those who talk in terms of *national universities* and *national sciences* to consider, particularly in post state-socialist Eastern Europe and, very particularly, before they ask universities to take care of themselves. "The ideological content of the knowledge produced in the University is increasingly indifferent to its functioning as a bureaucratic enterprise; the only proviso is that such radical knowledge fit into the cycle of production, exchange, and consumption. Produce what knowledge you like, only produce more of it, so that the system can speculate on knowledge differentials, can profit from the accumulation of intellectual capital."[13] One might still wish to question Clark's approach in presenting the five entrepreneurial universities as heroes. They have done what they have done to ensure their own survival as globalization—which for Readings, in the case of higher education, equals Americanization—continues in European higher education. With the exception of a small number of highly elitist schools, it is the only way for mass higher education to survive. As Readings argued, this is not about politics, this is not even about culture, this is about money: "Global 'Americanization' today (unlike during the period of the Cold War, Korea, and Vietnam) does not mean American national predominance but a global realisation of the contentlessness of the American national idea, which shares the emptiness of the cash-nexus and of excellence."[14]

The East European Entrepreneurial University

Nowadays, East European higher education is, by and large, more entrepreneurial than its Western European counterpart. Its driving mechanisms, however, are rather different. East European higher education has become entrepreneurial despite normative regulations and a culture strongly opposed to it. In most East European countries, either constitutional law or specific higher education legislation stipulates that higher education is provided free of charge. Nevertheless, even large public universities like the Chuvash State University in Cheboksary, at the heart of Russia, generate up to 80 percent of their revenue from tuition fees, quite apart from some 1,500 private higher education institutions that generally receive no public funding.

Entrepreneurship in East European higher education is a very narrow concept that basically involves charging students a fee or several fees, often in a rather irregular manner. This assumes two institutional forms. Initially, the new private higher education institutions, established after 1989, having no other source of revenue, started charging students a fee. Soon afterward public universities followed the same path as the funds promised by governments took too long to arrive.

Issues related to private higher education in Eastern Europe have been, if not widely then certainly aired over the past decade, usually from a critical perspective. Reports by international organizations, for example the Council of Europe, as well as more scholarly writings accuse these institutions of providing education of dubious quality, not having full-time faculty, not having proper infrastructure, even of (as in the Romanian case) copying public university curricula, and so on and so forth. This critique, often justified, is only a part of a broader picture. The detailed programs are sometimes ignored, for example where private universities are forced to copy *normal* universities to receive accreditation. If one compares the new private universities with the public institutions that the post-communist states have great difficulty in financing, then one can easily see that the general situation in public universities is not much better. Their infrastructure is falling apart. Funds for books and journals are unavailable. In some places, universities are closed for several weeks or even months during the winter because of the inability to pay the heating bills. Phones are disconnected. And the faculty is ageing. For example the vice-rector of the Comenius University in Bratislava, Slovakia recently reported that the average faculty age in his university was 63 years.[15] As private universities employ public university faculty on a contract basis, usually the more dynamic among them, there is no reason why their quality should be markedly worse than the public universities, which with a few exceptions are in a disastrous state.

The most glaring error the critics of East European private universities make is in considering them to be independent higher education institutions. Often they are not. They are in many important respects related to traditional public sector universities and constitute what Clark calls the *developmental periphery*. These institutions emerged in the early 1990s as independent entities because rigidity in public sector universities did not allow the creation of the new programs that emerging markets required (e.g. finance, business administration, etc.). Furthermore, they were unable to meet the faculty's expectation for remuneration. The faculty who could provide the courses the market demanded found physical space to meet this demand and earned the additional income without which many of them would probably have left the educational sector. Had public higher education been more flexible at the very beginning, it could well have accommodated a large share of the demand that emerged after 1989. Only after several years did public universities understand that the inability to be flexible had caused significant losses, which sometimes bizarre methods, for example accreditation, are used to make good. The explicit purpose behind the introduction of Western methods of higher education management is nothing less than the closing down of private institutions and taking over of their fee-paying students.

Since the mid-1990s, public universities began to develop their own entrepreneurial profiles despite major difficulties this involved. In some countries, formally committed to upholding a free-of-charge policy, the initial step taken by public universities was to create their own private parallel structures. Thus, for example almost all of the 200 public universities in Ukraine have *their own private universities*.[16] In effect, the rector of a public university is also the rector of a parallel private university that operates on the same premises and controls the money collected from private students who share the same classroom with public university students. Similar reports have come from Russia. It has been reported that in Hungary departments often run private limited companies under the same name as the departments themselves.

Gradually, an understanding developed in East European countries that without private cost-sharing, running their massive higher education systems is impossible. Entrepreneurial universities in Eastern Europe usually operate on the brink of survival. The money that can be collected locally allows no extravagance. The case study presented later, however, differs. In many respects, it is quite unlike a traditional East European entrepreneurial university. Yet, it does not resemble any of its Western counterparts, either. It is well organized and, in a way, highly entrepreneurial and successful. It does not, however, share any of the values that underlie the idea of an entrepreneurial university. Ultimately, it is difficult to say whether such an institution has education as its primary mission, or whether it is dominated by other goals.

The Story of the Academia Istropolitana Nova

The Academia Istropolitana Nova (AINova) is entrepreneurial in its own way, and perhaps even more importantly, highly adapted to the opportunities present in the post-communist environment. The AINova has appeared among the finalists of the Hannah Arendt Prize competition three times.[17] Although it has never won the prize, the very fact of its having reached the final stage on three occasions during the five years when the prize was awarded, carries a clear message: for the international higher education community the AINova appears as one of the most impressive reform initiatives in East European post-communist higher education. The AINova is by no means an exemplary entrepreneurial university and in many respects it remains unique. Examining its uniqueness is, however, helpful in defining the possible limits of the entrepreneurial response in higher education—East, as well as West. As this case study suggests, the range of entrepreneurial adaptations in higher education is almost infinite, rendering Clark's theory less conclusive than the management gurus would wish it to be.

The AINova is a postgraduate school offering one-year professional programs in fields that border with applied social sciences. However, the mix of the programs does not indicate any particular logic beyond the fact that most programs were not widely on offer in 1991. The school is located in the small town of Svjaty Jur [Saint George] a few kilometers from Bratislava, the capital of the Slovak Republic. The history of the AINova can be traced back to the days immediately after the 1989 Velvet Revolution in what was then Czechoslovakia. Although there was an institution of higher learning called Academia Istropolitana (AI) operating in Bratislava back in the fifteenth century, only the wildest imaginings could possibly connect the two institutions. The school faced many challenges during the decade that separates us from the first discussions about founding a new higher education institution free from communist ideology and providing advanced training in fields of studies that the communist regime was not particularly keen to promote. It is still too early to say whether it will be able to extend its operations beyond the *transition* in Slovakia.

The history of the AI (it was transformed into the AINova in 1996 under pressure from Mr. Mečiar's conservative government) started in early 1990 with discussions between an international financier, the philanthropist George Soros, and some Central European intellectuals related to the *Institut für die Wissenschaften vom Menschen* [The Institute for Human Sciences] in Vienna. Given the political context, the involvement of Vaclav Havel in the process seems to be highly probable. While available documentation does not reflect much of the content of the negotiations, the AI report from 1996 indicates that in May 1990 a decision was reached to launch the Central European University in Bratislava.[18] The chosen location is particularly significant as it is literally located at the crossing of the borders of three countries—(Czecho-) Slovakia, Hungary, and Austria. One month later George Soros and the then deputy prime minister of Czechoslovakia, Mr. Hromadka, signed a joint announcement concerning the establishment of the Central European University (CEU) with campuses in Prague and Bratislava.[19] Roger Geiger in his study on single donor universities noted, "The Central European University was created by financier George Soros in 1991 to compensate for areas of higher education that were neglected under the previous Communist régimes."[20]

In November 1990, the AI was formally launched in Bratislava as a *preparatory project* for the CEU. While the ministers of education of three Visegrad countries (Czechoslovakia, Poland, and Hungary) and Austria endorsed the AI project at their meeting in Vienna in January 1991, only a month later the project took a different turn. For unknown reasons, but most probably related to responsibilities to be assumed by the Slovak side, the AINova was excluded from the CEU project by George Soros. While the CEU continued to prepare for the admission of the first student cohorts in Prague and Budapest, the AINova was chartered by the Slovak legislators as a public higher education institution. It was granted a *semi-budgetary* status that authorized the school not to follow the structure of public universities and all the related faculty, staff, program, and financing requirements.

By spring 1991, the AI entered its first stable period of activities. Funded by the Slovak government it began by offering short courses and public lectures whilst simultaneously preparing to launch regular programs. The aim of this seems to have been to become a center for advanced studies and the development of a strong academic core and research capacity. In July 1992, the entrance exams for the first cohort of students were held and in September 72 students (68 Slovaks and four foreigners) started their studies under three programs, Public Administration, Environmental Planning and Management, and Architectural and Urban Conservation. In 1994, a two-year economics program, the Professional Program in Applied Economics, was introduced and preparations began to launch a fifth program—European Studies. By the 1995/96 academic year, the student body had grown to 80. Most of the training, however, was still provided by visiting faculty. AI's own faculty amounted to 12 people in 1995/96 while 78 visiting faculty, two-thirds of them foreigners, carried the main burden of teaching.

By 1995, the AI academic programs seem to have reached a certain stability. At the same time, political problems continued to gather around this institution that neither fitted the traditions nor practices of the post-communist environment. Although the AI's founding director was a former deputy minister for higher education of the Slovak republic,[21] her standing could not prevent open political conflict, which with changing governments changing, her rather independent personality might well have generated. Whatever the reasons, the Slovak authorities wanted a more regular role for this irregular institution within the national higher education system and to control it. Eventually the conflict reached an international scale with the foreign community, beginning with Lord Dahrendorf supporting the AI on the one side and Slovak leadership ranged on the other.

The Slovak government wanted to incorporate the AI into the national higher education system but did not know exactly how to do so. The minister of education of the Slovak republic issued a decree to transform the AI into a *unit* of Comenius University, Bratislava. Since the practical steps had not been thought through, this decision remained a dead letter. The Slovak Ministry of Education continued to cut the AI budget, forcing it to find other funding sources. In criticizing the Slovak government one should note that 3,000 dollars of public funds per student, perhaps five times more than was available to other universities, posed a serious issue in the Slovakia of 1995. As the whole issue moved into the political arena, foreign funds seemed to be readily available and covered almost two-thirds of the AINova budget in 1995. The AI became a political issue between the pro-reform and conservative forces to the extent that the police attacked a group of European Studies short-course participants.[22] Political persecution, however, backfired. Every new attack increased its international standing, as the last bastion of resistance against an authoritarian regime.

A major crisis developed by the end of the academic year 1995/96. The AI leadership found it impossible to continue working—formally as a higher education institution, but under imposed and direct control of the conservative government under Prime Minister Mečiar. At this point a radical decision was reached by the AI's exclusively female leadership.[23] To abandon the AI, its building, facilities, library, everything and set up a new institution elsewhere. During the academic year 1996/97 teaching was suspended, and in autumn 1997 the AINova reopened Svjaty Jur, as a civic association with its students and faculty defined in the bylaws as *members of the association*. This rather unusual transformation (and the continuity it provided) was suggested by three large international foundations providing a rescue package, which allowed the AINova to rent and renovate facilities, establish a new library, and resume its activities. An alternative explanation might be that, as government funding ceased the only rational solution was to generate political capital out of a most dramatic exodus.

Having come through this change, the next major shock came in 1998 when the AINova director, Dr. Alena Brunovska died. Early in 1999, the new director came into office. She brought experience as a lecturer of journalism at the Comenius University in Bratislava and administrative experience as the executive director of the Slovak Foundation of Independent Journalism. The opening of a one-year postgraduate program in journalism in January 2000 and the successful renegotiation of the support of the three major foundations were the most important results of her first year in office.

Governance

The AINova has two levels of governance. The institution is governed by a Board of Trustees. Each of the programs has its own Advisory Board. The Board of Trustees (BT), a seven-member international body includes representatives from academia, business, and the nonprofit sector. In 1997–2001 it was chaired by a former U.S. diplomat. The BT meets two or three times a year and oversees the overall activities of the AINova. It follows the life of the school and its financial situation very closely. To do this it receives detailed reports from the director's office. The BT also decides about opening new programs, follows annual student recruitment campaigns, and is informed about problems occurring both at the institutional and program level. The Program Advisory Boards oversee each program's academic activities and make program-level policy and administrative decisions. They develop and approve the content of the study programs, and usually include the leading faculty of the respective programs as well as external representatives.

The AINova governing structure is remarkable in East European higher education. Though some countries recently attempted to introduce the BT to the governing structures of universities, they have usually been rejected as an unjustified limitation to academic autonomy. Where such structures exist their role vis-à-vis the rector and the Ministry of Education is not well defined. This does not pose a problem for the AINova, since there is no external body above the BT to which the institution is accountable. Operating outside the formal educational system of the Slovak republic, it does not report to the Ministry of Education. This is not an unalloyed benefit. AINova's qualifications awarded in the home country or anywhere else, are not recognized nor is it funded by the Slovak government.

The size of the governing structure gives one board member (trustee or advisory) for every two students. Since each of the programs have their own supervisory boards with significant powers on both academic and administrative matters, the institution itself resembles, at best, a confederation with only limited authority delegated to the top management and the BT. This division is further stressed by the fact that the programs raise funds and use them as agreed between the program and donor with the institutional management having only limited role in the process. The BT has provided all the information, but its power to intervene in program-level activities is limited.

Funding

Over the past decade, the AINova moved through a whole range of funding patterns. It started out with full support from the Slovak government. By 1995, a governmental funding had been cut to approximately 36 percent of the total budget. The change from AI to AINova involved 100 percent funding, provided by foreign governments and foundations.

The AINova funding pattern and the philosophy behind it is unique, particularly for an entrepreneurial higher education institution. The leadership of the AINova strongly believes that higher education is wholly a public good and should be provided to students without charge, no matter what. Despite the difficulties experienced it has continuously rejected proposals to charge students a fee. Currently, none of the regular AINova students pay a fee. For foreign students from other East European countries as far distant as Central Asia, the AINova covers their travel to Bratislava. All students receive stipends significantly higher than the minimum salary as defined by the Slovak government to cover living expenses in Slovakia. The AINova budget in 1998 was composed of 98.2 percent contributions by foreign governments (particularly Austrian) and private foundations (figure 7.1).

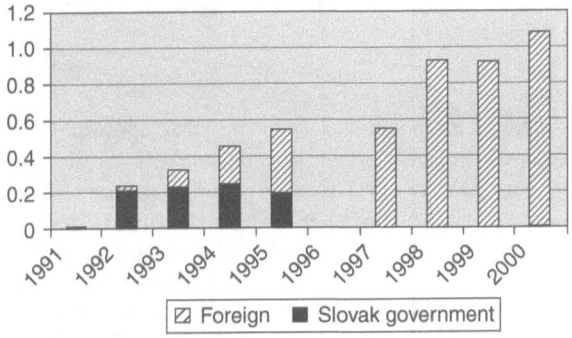

Figure 7.1 Dynamics of the AINova funding (1991–2000).

The remaining 1.8 percent was generated through short programs, particularly in English-language training.

Recently, concerns seem to be developing that the AINova cannot continue with such a funding scheme for very long. Yet, the AINova has benefited immensely from the presence of Mr. Mečiar's government. As the only independent higher educational institution in Slovakia, generous foreign funding was made available, although, after the change of government in 1999 and with Slovakia included among the EU accession countries and invited to join the OECD this may soon end. Currently, the AINova spends at least ten times more per student than any other higher education institution in that country. The monies spent are determined less by realist cost than by the amounts foreign donors are willing to pay. Because of particular political concerns donors have been generous until recently. With their generosity they have, however, inflated the operating costs of the institution so that it has no chance of surviving in the local economy should donor funding discontinue. It is hardly possible that students themselves would be willing to invest in nonrecognized education at a level comparable to that of the foreign donors. Even some of the most elite private universities in Eastern Europe that generate their budgets completely from student fees do not charge more than 3,000 dollars per annum. That leaves the AINova dependent on the donor funding, which has been drastically reduced in 2002 and 2003. As by Spring 2003 AINova has no longer been able to pay the rent, it has been requested to vacate the premises. Even if the government decides to enter and support the AINova on a scale comparable to other Slovak universities, it would only be a fraction of what the school has grown accustomed to spending.

Recently, the AINova began to offer short, fee-paying professional courses, for example in European Law, to members of the Slovak Parliament. Even if these short, professional courses generate up to 10 percent of the school's budget,[24] the rest remains uncovered. One solution could be to revise the school's underlying philosophy, the cost of its governing structure, overhaul the current practice of using a large number of visiting foreign lecturers, as well as modifying its academic profile. If this is done, the AINova will most probably lose some of its unique characteristics that set it apart from East European independent education, among which, following the infamous old communist maxim—has been the distribution to all according to their ever-growing needs. It is indeed ironic that generous financial assistance from *capitalist* countries has allowed such an island to be created in Eastern Europe, perhaps to fulfil some of the old dreams.

Academic Programs
The AINova's academic profile still cleaves to the initial ideas of 1990–91, that is, providing training on an eclectic range of professional fields that the communist regimes in Eastern Europe were not particularly keen to develop. The negative logic on which the academic core of the school was established is not, up to creating a strong higher education institution. A decade later, the initial idea of becoming a center of excellence is no closer than it ever was. The AINova has not developed a core competence in any of the fields it teaches. Its faculty and staff are some 30 people, filling largely administrative positions while the teaching faculty is made up of foreign and, to some extent, local visitors. While the AINova developed certain

research activities in 1993–95, with the major change in 1996 those attempts seem to have foundered.

Currently, the AINova academic program remains rather eclectic and includes: Environmental Policy, Architectural Conservation, Applied Economics, European Studies, and Journalism. The AINova defines itself in its charter as an institution providing advanced programs in Social Sciences and Humanities. Sometimes it presents itself as an institution providing *liberal* education. While the general identity is not very precise, humanities are not greatly apparent, either. One could possibly argue that a strong administration without an established academic core is a particular benefit in a rapidly changing environment, making it possible to react quickly to emerging market needs, providing a meeting point for visiting faculty and needy professionals. The only new program that the AINova has developed over the past five years is in journalism. This, however, is less related to any new need in Slovakia or Eastern Europe but rather to the professional background and interest of the new director. Distribution of powers within the AINova, particularly the power concentrated at the level of programs, makes it virtually impossible to discontinue a program so long as it has funding available from foreign donors. The level to which donors perceive Slovakian and East European needs for training remains, however, open to question.

Rather than being beneficial, the current structure has become a serious hindrance for the school. The way it operates prevents it from responding to needs of Slovakia or Eastern Europe. Instead, it responds to what foreign donors think those needs are, if that. Only the funding agencies directly have the necessary power to produce *market signals* for the AINova. So far, they have chosen not to do so. The current situation could be seen as one in which donors invest in personalities and broad intentions and less in particular educational programs. From this standpoint the AINova's educational philosophy can be interpreted as a particular form of learned helplessness. Although foreign donors often talk about "capacity building" in Eastern Europe, in the case of the AINova, no capacity has been built. Everything donors provide is immediately consumed by local administrators and foreign faculty.

Lacking an academic core significantly limits the AINova's chances of academic recognition. In the Slovak environment, there are several reasons why it does not qualify for state accreditation: first, it provides only postgraduate training and does not qualify for the title of university; second, to be accredited, the AINova would need to hire a number of full professors with full-time commitment. The AINova's own argument is that the school cannot afford this because of the high cost.[25] Two counterarguments are at this point: first, public higher education can meet these requirements and do see without the means available to the AINova; second, visiting foreign faculty may not be any cheaper than hiring local or East European young Ph.D. holders who have recently completed their studies abroad. However, young and qualified faculty may choose not to commit to an institution that is bereft of legitimate academic status. Thus, the AINova is in a situation similar to that in East European, entrepreneurial higher education in general: faculty receives its pay from the private sector, and its status and job security from traditional public universities.

That AINova can maintain the status quo is due to the complete freedom that the current situation provides—it is not actually accountable to anybody. As long as it offers totally subsidized education to local students and those from the less-developed regions in Eastern Europe and the former Soviet Union, the status of its credentials will hardly ever be raised. Arguably, it is the material benefits the AINova offers rather than education that attracts students to Svjaty Jur from as far afield as Kazakhstan. The situation would, however, change dramatically were students to be obliged to cover any significant part of their tuition cost themselves. Clearly, at that point, the school will face major recruitment difficulties. The AINova's formal strategy is to seek program accreditation from Western partner universities. This, however, may have serious cost implications and time factors, and still might not solve the core problem—lack of academic capacity. Certainly, an entrepreneurial Western partner university could grant recognition in exchange for a cash payment. Hopefully, we have not reached that point yet, although the situation is changing fast and *digital diploma mills*, even accredited ones, are on the rise.

The AINova in the Context of Entrepreneurial University Theory

Finally, it would be interesting to see how the AINova fits into the theory of the entrepreneurial university, as developed by Burton Clark. The AINova is not an interesting case on account of its being an exemplary case of an entrepreneurial university. The AINova is in many respects a unique institution in Eastern Europe, representing neither of the two big categories of entrepreneurial higher education: new private institutions moving flexibly in the market and earning their revenues by charging fees to students; and traditional public universities attracting fee-paying students with their official status. In a sense, the AINova profits from the delivery of highly subsidized services the cost of which is far beyond what the local market is ready to pay. It has an artificially established market niche for this, and can exist only as long as subsidies are available.

Within the perspective of Clark's main conclusions on the entrepreneurial university the AINova is not representative, either. The first characteristic of the entrepreneurial university is *the strengthened steering core*; one should admit that the steering core is perhaps the strongest element of the AINova. It has played a particularly important role in leading the university through the crises it experienced during its short history, particularly in 1996 when it was able to move the whole institution elsewhere, although this aspect should not be overestimated. Moving the former AINova from Bratislava to Svjaty Jur meant moving fewer than 30 people. And as for the strengthened steering core it is not exaggerated to point out that we are dealing with a single person.

The second common characteristic of the entrepreneurial university turns around the *enhanced developmental periphery*. Earlier, we pointed out that one of the main characteristics of many East European private higher education institutions is their organizational weakness. They are better seen not as independent institutions, but as a developmental periphery of traditional universities. Even though the AINova, because of the political pressures it faced and the need to develop a clear alternative identity, has a stronger organizational structure than many other universities in the region, still it largely operates on the periphery of traditional universities. It offers opportunities to local and to foreign visiting faculty to develop new courses and gain experience within a different setting. One cannot really apply the concept of "the developmental periphery" to the AINova because it constitutes the developmental periphery of those higher education institutions whose faculty teaches there.

The third characteristic Clark identifies with the entrepreneurial university is *the discretionary funding base*. In this respect, the AINova's funding offers an interesting variant. While the entrepreneurial universities Clark analyzed usually diversify in their funding base as governmental support becomes scarce, the AINova budget almost doubled soon after it lost the support of the Slovak government. Interviewed by The Chronicle of Higher Education, the director of AINova claimed that the loss of partial state funding was a small price to pay for complete freedom and that remaining independent would be *healthier*.[26] Still, the AINova faces a strong need to diversify its funding sources. Yet, even if it is able to follow other private higher education institutions and offer programs for which the population is willing to pay, it faces the need to reduce its operational costs to levels comparable to those of 1994 or 1995, cutting it approximately by half.

Stimulated heartland is, Clark argued, the fourth key characteristic of the entrepreneurial university. The principal problem the AINova faces these days is that it does not have any academic heartland—all it has is management. It is faced with serious problems in receiving recognition for its programs because of its lack of inbuilt academic capacity. In principle, a structure like the AINova might prove successful in changing times by providing the market with different educational services. While some progress has been made, it seems its regular program is frozen and does not reflect change in society. A recent externally commissioned report on AINova's future prospects highlighted the need to establish clear criteria for discontinuing programs.[27] This may be a sign that academics, particularly the part-time faculty that offers its services at the AINova, has been able to gain indirect control over management and is in a position to generate relative stability to its own benefit, whilst opposing changes that could benefit the entire school.

Finally, Clark stresses *entrepreneurial belief* and explains how important entrepreneurial culture is in leading an institution through difficult organizational changes that crystallize finally as a *saga*. The AINova is a controversial institution—entrepreneurial without being entrepreneurial. Perhaps it was never meant to be entrepreneurial, its leaders never studied the educational needs of changing society, or responded to anything but their own dreams. Initially, the public interest generated within the context of launching the CEU provided it with access to public and some private foreign funding sources. Later, becoming symbolized as the only independent higher education institution under a politically hostile regime, foreign support was almost guaranteed. Somehow, during certain period not being entrepreneurial proved to be the best entrepreneurial strategy. It provided access to customers of a much higher caliber than local students.

Conclusions

The AINova offers an appropriate case against which to test the limits of recent entrepreneurial university theories and models. Those more accustomed to the tricks of Marxist dialectics, for example quantities becoming qualities, structures overcoming their own limits, and so on, would have an almost infinite field of study here. Instead of selling its services to consumers, it has found foreign donors who cover the cost of running the institution and do so in abundance. Thus, the AINova distributes its services for free and has created an artificial demand for it. Despite all the benefits of such a scheme, it is by no means sustainable. One may speculate that it provides a pretext under which some of the bad old habits, particularly inefficiency, easily continue to flourish. Until recently, one of the most conservative East European regimes justified foreign agencies and governments supporting the school. This justification no longer exists.

The second significant characteristic of the AINova is that, it has turned our understanding of how an academic institution works upside down. Instead of having an academic core and different services related to it, the AINova has developed strong administrative core, at least in relative terms, having one administrator for every three students. Academic elements related to it seem to remain relatively weak with part-time visiting faculty—local and foreign. Academic activities are a clearly peripheral function. Recently, the school seems to have set its mind to thinking about a possible change, but without any practical results so far.

The AINova has come through a decade of major change in Slovakia and Eastern Europe. It set out with billowing idealistic notions of training new leaders for a new society. Slovak society decided otherwise and for some years moved backward. This forced the withdrawal of the AINova from the educational sector and its becoming *alternative* to the extent possible given the prevailing conditions. They allowed it to rally necessary support behind the school. But, it also developed habits that cannot be continued.

Chapter 8

When East Meets West: Decontextualizing the Quality of East European Higher Education[1]

> Therefore, when the way is lost, only then do we have virtue;
> When virtue is lost, only then do we have humanity;
> When humanity is lost, only then do we have righteousness;
> And when righteousness is lost only then do we have propriety.
>
> Lao-tzu[2]

And when universities lose their intellectual commitment only then do we have quality.

The East European Quality Movement

One of the most common characteristics of the higher education systems of many post state-socialist countries is the great attention paid to quality assurance. Since 1989 higher education quality agencies have been established in Central Europe as well as in many of the former Soviet Union countries. Romania initiated the *quality assurance* movement in the East in the early 1990s. More recently, Armenia's Education Act introduced the concept of accreditation in this much-troubled Caucasian country, apparently as a part of its Westernization program.

It is symptomatic of post-totalitarian higher education that quality assurance is uniquely addressed through the process of nonvoluntary accreditation and that quality assurance and accreditation are considered synonymous. The latter has given many East European quality experts a good reason to argue about the total lack of higher education quality assurance under the former political regime,[3] although this may not be a true reflection of the situation. Whilst it is true that there were no accreditation procedures as such, there was actually no need for them because the quality of higher education was assured through more direct means such as direct control over content, as well as the administration of higher education. As Ratcliff put it, "it should be noted that accreditation is not necessary if all institutions ascribe to identical standards for admission and graduation."[4] Accreditation only became meaningful when the state lost control over a noticeable part of higher education. It did not usually happen through an increase of the distance between the state and the traditional higher education sector, but through the emergence of a significant number of new private higher education institutions.

The massive rise in private higher education institutions came to many of the East European governments as a great surprise and it took several years for legislation to regulate their operations to be adopted. Sterian presents the uncontrolled and unregulated growth in the number of higher education institutions as the main reason for launching the system of quality assurance in Romanian higher education.[5] Within five years of the December 1989 revolution, the number of universities in Romania had grown from 44 to 130. In all, 74 out of 86 new universities were private. In many countries the proportion and speed of growth in higher education has been similar. For example Estonia had six higher education institutions in 1989 and 48 in 2000. A common reaction to this has been to introduce quality assurance measures. In time of great interest in the quality of higher education in Europe, it provides, amongst other things, a convenient cover for almost any policy agenda a particular government may have.

Ratcliff summarizes the dominant quality assurance models:

> ...the French model of external reviews serves as an archetype for quality of the educational program or institution through implicit or explicit comparisons. The English model of quality assurance through peer review serves well the aim of enhancing program effectiveness and improving teaching and learning. The American model of voluntary accreditation, drawing on both traditions, ensures that the quality review process is conducted outside the context of government funding and control. Self-study criteria and peers conducting reviews are promulgated across state boundaries insuring that one system is examined within the context and experience of other but with the primary aim of program enhancement and improvement of student learning.[6]

It is ironic that instead of following the largely external French model, which seems to fit East European thinking and policy discourse better, many countries have chosen the Dutch model. Jones and Ratcliff describe the origins of the latter as follows: "Beginning in the mid-1980s, the Netherlands developed new policies for program review and quality assurance adapted primarily from the United States model of self-study, peer review, and self-regulation, giving particular attention to the quality of teaching and instruction."[7] According to Frederiks and colleagues, "Quality management has been an issue in Dutch higher education since the 1980s. Quality Assessment of education was introduced on the political agenda as a part of the new policy of the government, with the policy paper Higher Education: Autonomy and Quality (1985). In exchange for a larger measure of administrative autonomy, the universities promised to retain and enhance their levels of quality in education. Quality assessment then appeared on a systematic level and nationwide scale in 1988, when the Association of Universities in the Netherlands (VSNU) implemented its new responsibility."[8]

East European emergent democracies are not shy about presenting their success in Westernizing higher education. The launching of quality assurance mechanisms is often presented as one of the significant examples of catching up with the West. There is, however, a significant gap between the Dutch model of steering from a distance and East European policy of *direct* administrative interference. Unique unanimity prevails among East European countries on the issue that higher education institutions that do not pass the state accreditation should be closed by administrative means. The experience of countries where the new, so-called Western quality assurance methods have been applied for long enough to expect real outcomes, for example Romania, demonstrate a high but selective productivity of the process. Many, if not the majority, of private institutions are closed down while the public institutions remain intact.

Ten years earlier, many expected the opposite to happen, that a communist higher education establishment fatally contaminated with narrow careerism and brainwashing would be closed and an alternative based on free intellectual inquiry would be established. Following the critique of the late 1980s, it would be difficult to define the precise quality that is so much higher in the traditional higher education and provide a good reason for why the new private sector is performing so much worse. Both sectors use the same teachers, many of them working part-time in three or four universities, public and private.[9] The argument about the poor infrastructure of private universities also fails because many of the governments have not been able, for almost a decade, to invest in the infrastructure of public universities either. The blame private universities bear is that the majority of them, but not all, have been unable to offer better higher education than the deteriorating public universities. In practice, instead of leading higher education toward improvement, the quality agencies seem to have taken the role of the "quality police" protecting the monopoly of traditional institutions. This appears to be comparable to what Broadfoot says of the impact of the current quality measures on a changing university, "It is therefore peculiarly ironic, . . . , that at the very time when these same institutions are facing a challenge to change on what is arguably an unprecedented scale, they are being caught up as never before in the trammels of formal assessment which are binding them ever more tightly to the status quo."[10]

The reason for unsuccessful accreditation of private institutions may lie in the policy that they are sent to the accreditation fire first while state-run universities follow later, if at all. Traditional, politically powerful universities have the channels to negotiate accreditation decisions even if the formal requirements are clearly not met. On the other hand, given the dependence of the state higher education institutions on the state, the latter is to be blamed if things go awry. Universities in Russia, for example do not wait for too long to remind the state of its unmet funding commitments that may amount to half of the allocated budgets.[11] When the state starts accrediting its universities under these circumstances, it is actually accrediting itself.

Two Models of Accreditation in Eastern Europe

We can broadly distinguish between two models of accreditation in the region of Central East Europe and the former Soviet Union. More Western-oriented countries like Hungary, Romania, and the Baltic States of the former Soviet Union largely follow what is perceived as the Dutch model. The Council of Europe has done significant work in importing this model into Eastern Europe, a process that usually combines self-study and peer-review. As a major difference from the Dutch approach the Ministry of Education has ownership of the process directly or through a pseudo-independent agency. They support the new initiatives with statements like "You are doing all the necessary things, and you are doing them in a right way."[12]

Western experts ignore how an initially formative evaluation model aimed at program improvement has been turned into a summative one guiding rather significant administrative decisions concerning, among other things, the continued existence of an institution. Mentioning voluntary accreditation, as the author of the statement quoted earlier did in his presentation to the Romanian quality agency (NCAEA), seems particularly hypocritical, if not cynical, in a context where demonstrating insufficient volunteerism leads to condign punishment. However, there are experts, even in countries with older democratic traditions, who believe the problem can easily be solved by threatening even worse punishments: "The thrust of the argument is that by heavily penalizing any visited departments which are found to be cheating in their self-assessment, honest revelation can be induced by reducing to below unity (indeed to close to zero) the probability of a visit."[13] The methods this particular expert expects the British quality assurance mechanism to apply are not so very far from the "self-criticism" campaigns organized earlier in the twentieth century in some other countries. Lavrenti Beria, the infamous Soviet internal security chief, would be delighted to read about it. During the purge of 1937, Comrade Mezhlauk, a member of the Central Committee of the Soviet Communist Party, advised the *rightist conspirators* against the communist Soviet Union to confess in the following way, "I am a viper and I ask Soviet power to exterminate me as a viper."[14] What some British quality theoreticians expect to see in the self-assessment reports would not be so greatly different. To meet the expectations in British higher education policy circles a report might read, "we are a low quality university and we ask the Higher Education Funding Council to cut our funding."

The final decision on accreditation belongs usually to the executive branch of the government. In Estonia the Minister of Education approves the proposal made by the Higher Education Evaluation Council.[15] In at least one country, Romania, the final decision concerning program, as well as institutional accreditation belongs to Parliament.[16]

The Russian Federation has developed another quality assurance model that served as a prototype for many of the former Soviet Union countries, particularly for the newly independent states of Central Asia and Trans-Caucasius. Its outcomes are, however, not very different from its Central European counterparts—limiting the role of private provision of higher education and dictating the content of studies. All the countries concerned initiated quality assurance procedures after expansion of the system had been going on for several years. This implies that a negative accreditation decision does not mean a new institution will not be established but rather one that is currently operating, will be closed.

In the Russian quality assurance process, accreditation constitutes the last of three tiers.[17] First, in order to operate, every higher education institution should have a government license. Again, before this regulation was passed hundreds of institutions had already been established. At one point all these institutions had to apply for a license retrospectively. Prior to issuing the license, teaching and conditions in the applicant higher education institution and the qualifications of its teaching faculty are checked.

The second step is attestation, which means state examination of graduates, administered by the state attestation services. For an institution to be evaluated positively 50 percent of students in three consecutive cohorts should pass these examinations. Attestation is renewed once every five years. Positive attestation seems to be a necessary but not sufficient condition for accreditation. Teaching conditions, administrative procedures, and other aspects are studied as well. For example in Belarus the final phase of the accreditation is a month-long site visit by an expert commission set up by the Ministry of Education. When the commission finds that an institution has complied with all requirements, the institution will finally be accredited.

Accreditation confirms formal equality between a leading state higher education institution whose program in a particular field of study has been adopted as the standard and an accredited institution. It is done, however, in an extremely narrow way—on the basis of a near-complete match between the contents of study, teaching methods, and results of two programs. This may raise questions about the purpose of alternative higher education if all alternative elements are banned. The answer may well be that there is no such purpose. Whether this reveals systematic discrimination against private initiative in higher education or the belief of leading academics that they possess the ultimate truth needs to be studied further. Quality assurance has been introduced as a part of the politically driven thrust to westernize and applied in the only possible direction. As the process is controlled by senior academics it slows down rather than speeds up renewing the content of higher education studies in Eastern Europe.

Defining Quality Standards
The end of the Cold War has, in addition to shedding new light on the military and economic strength of some of the formerly dominant countries of Eastern Europe, had a similar impact on the reputation of their educational systems. As the fictitious nature of their economic power and military strength has been humiliatingly revealed, education that had been recently praised for launching its scientific and technological foundations has been, to some extent, reassessed by the international community. Recent studies clearly indicate the pain of reconciling the official discourse of the unprecedented development and success under communist rule since 1917 with the undeniable misery of higher education in the 1990s.[18]

Political and economic conditions created insurmountable problems for systems of quality assurance. A decade ago, East European higher education was blamed, not always unjustly, for serving the communist establishment. Now, the new systems have failed to address many of the important reform issues. They have channeled their reform into cheap but cosmetic changes. What post–Cold War East European higher education needs, if it is to control the external forces that push it toward change that many of its organizations may not be able to meet, is an abstract and distant quality standard. "Avoid major upheaval" has become a universal, though relatively unproductive policy imperative.

Ratcliff noted, "Central to the process of quality assurance is how and who gets to decide what constitutes a quality."[19] Elsewhere Jones and Ratcliff complain, "Unfortunately, there is no absolute agreement to what constitutes quality...."[20] Not everybody shares this view. There are people who know exactly what constitutes quality. But there are also those who believe there is no need to define it precisely.

The dominant thinking among the theoreticians on quality of higher education appears to be based on the premise that quality is highly contextual. Brennan and colleagues express this by defining the quality of higher education in terms of its fitness for the purpose.[21] Hardly surprising then, that the new European quality movement has found this definition extremely useful. It allows technocrats controlling the quality process to operate at an abstract level without having directly to deal with the substance of study. Hence, quality is assured by a simple comparison between expectations and real outcomes. Possible difficulties arising from the unquantifiability of some factors may not be considered a very serious issue. It is probably assumed that combining the process with peer-reviews resolves all outstanding concerns, although this is not necessarily so. Peers may or may not have a margin of freedom and ability to reach beyond the semiautomatic process of applying the quality formula.

Technocratic processes, amongst other things, allow one to deal with the concerns of those ever-complaining intellectuals who believe that a statement on absolute relativism (just define the purpose!) is self-contradictory and who look for more stable reference points. Trying to reach universal applicability through evacuating the concept of quality is problematic.

> Quality... you know what it is, yet you don't know what it is. But that's self-contradictory. But some things **are** better than others, that is, they have more quality. But when you try to say what the quality is, apart from the things that have it, it all goes **poof**! There's nothing to talk about. But if you can't say what quality is, how do you know what it is, or how do you know that it even exists? If no one knows what it is, then for all practical purposes it does not exist at all. But for all practical purposes it really **does** exist.[22]

Absolute relativism as an approach to quality assurance may have economic justifications at a time when quality assurance has become just another global business—disappointing those who commission particular services and which may not fit the purpose of generating income. Its philosophical and moral roots are, however, weak.

One can easily imagine a situation where a college of applied biology or some similar vocational higher education institution is established in a not particularly democratic country and accreditation is sought from a Western entrepreneurial university. If it does not break a UN embargo and agreement is reached concerning the service fee, there seems to be nothing within the current quality discourse to prevent the college from being accredited. There seems to be no reason why the real purpose of the college (even if the latter is defined in terms of training high-quality specialists for a biological warfare program) should constitute an obstacle. While the desire to earn profit puts a strong pressure on universities to ignore other less tangible considerations, it is difficult to separate quality from fundamental values, for example from the value of human life. This brings us back to Pirsig's quality dilemma and Putnam's fact and value question,[23] often ignored within the current quality discourse as irrelevant. Bringing in entrepreneurial visiting peers may not maintain the vanishing values of the academy for too long. Precisely why Western liberal democracies have ignored certain value issues may well be the assumption of the universality of those values even though the rise of the entrepreneurial university exerts strong pressure upon them.

Fitness for purpose is not a good solution for East European countries, because there is apparently no purpose. So, a solution to the dilemma is to be found at the other extreme. The quality target is defined in the infinite distance. As a quality target for higher education "meeting international standards" occupies a significant place in the East European quality discourse. The former Eastern bloc countries in Central Europe and the former Soviet Union brandish meeting "world standards" as their main aim. They declare at every opportunity that these have actually been met. For example the State Committee of Higher Education of the Russian Federation prepared a special comparative report[24] to show that the Standard of Higher Professional Education fully met world standards as established by Cambridge, Princeton, Austin Texas, Bremen, Karlsruhe, and some other universities.[25] Even if there was any truth in the conclusions, whether providing such education within the Russian context is appropriate and compatible with the needs of the country was never broached. The East European quality discourse on the whole ignores the fact that if there is anything that can be called a "world" or "international" standard of higher education nowadays, it is closely related to training highly qualified labor force for a particular environment. It may well be that East Europeans do not want to have much to do with their own environment and circumstances (massive unemployment, unpredictability of the labor market demand for graduates, overproduction of graduates, particularly in science and engineering related fields). Why this should be, we will now examine.

The Inevitability of Absolute Standards in the Post-Totalitarian Society

Discussing higher education within the East European context has inherent dangers. First, if higher education in general is highly context-specific, then it would be easy to conclude that higher education under communist regimes was also context-specific. This raises the delicate issue about whether communist-era graduates are suitable for changed conditions. Such a view is politically unacceptable because it would leave not only many senior faculty, and also much of the new state nobility bereft of appropriate academic training.

Strong reasons exist and show why quality standards in East European higher education cannot be derived from the needs of the new economies and transformed societies. By studying the content of the education East European universities offer, clearly not very much has changed in comparison with the programs of 1989. Discontinuation of previously obligatory courses of orthodox Marxist–Leninist philosophy and Communist/Socialist Party histories may be the only large-scale changes. Even so the units responsible for teaching those subjects have been relabeled as departments of political science, sociology, or modern history. The same faculty now teach the new subject to the extent that they are able to.[26]

However, it would not be fair to blame the universities for not changing. Universities face a particular difficulty in trying to define their purpose in an era when, from a technical point of view, purpose is largely absent. Previously, these countries at least had formal mechanisms through which the developmental needs of the state-socialist economies were translated into university programs and the expected quantity of graduates. For example the structure of central planning in the Soviet Union collected information from enterprises on their anticipated need for graduates and, on that basis, prepared the student admission quota to higher education. Highly centralized structures were also established to prepare program and course prescriptions for each program and course. The post state-socialist transition has only broken a part of this process. The central level remains strong and often deals with drafting or adopting detailed program descriptions (the so-called state standards) for universities, as well as setting the *numerus clausus*. The feedback loop, through which the system of higher education was fed with information, has been cut and the structures of central planning responsible for it, dissolved. Somehow, one half of the former command system continues its rather independent life. While this situation could be justified politically, in the long term it undermines the credibility of higher education and prevents universities from defining their own mission and establishing direct links into society with its changing needs.

Universities continue to teach in a traditional way, retaining much of the old material, even though the relevance of the training may be questionable, though arguably the content of education in transition economies does not really matter. Universities still fulfil a very important social purpose, keeping a large number of young people and university staff busy in times when the situation in the labor market is particularly difficult, and do so at an extremely low cost. Recent discussions concerning, for example lifelong learning suggests that expanding the mission of higher education to accommodate growing numbers of the unemployed is a priority for post state-socialist economies. Various forms of higher education may soon fill the lives of many adults in developed economies as well, though it does not hold out much hope of entering regular employment before retirement and thus constitutes a formula for *eternal youth*.[27] One should not blame politicians in either the East or West if they find ways to make political capital from this situation. There are other interpretations beyond continuing progress and democratization. That they present their respective higher education systems as meeting universal standards is not surprising. Ultimately, this is what they are paid for.

Nevertheless, the post state-socialist countries in Eastern Europe and the OECD are moving in the same direction: toward quality decontextualization. While the thrust behind the OECD performance indicators' project is the commodification of knowledge on the increasingly global markets, East European quality discourse is based on different premises: first, that there is a final body of relatively stable knowledge; and second, that this body of knowledge is already at the disposal of the leading universities in particular countries. Leaving aside the possible provenance of this knowledge, whether copied from Oxford or established by representatives of a class-conscious proletariat, the framework itself implies the continuation of orthodox Leninist thinking. Groping for possible reasons behind the subterranean continuation of many ideas from communist ideology in contemporary Eastern Europe, one might conclude that the changes of the late 1980s were related less to a crisis of ideology, than to the manipulations by the second echelon of the functionaries themselves—the people who translated their former networks and positions into economic wealth. The ideology itself is not entirely extinguished.

The Multiplicity of Quality Contexts

That state-socialist higher education meets international standards rests on the achievements of Soviet higher education and that of its former East European allies in hard sciences and mathematics. Recent scholarship confirms that despite the inevitable need to invoke Marxist–Leninist rhetoric in introducing works on physics, scholarship and training in these fields is comparable with anything provided internationally. There is also little reason to make training in geometry context-dependent. The scholarly standards in the fields of fundamental research are established by the respective international research communities. However, whether the research community finds it acceptable (at least hypothetically) to be involved, for example in a peer-review of fundamental science serving the aims of global revolution of the proletariat or some similar aim is an ethical issue that only the community itself should resolve.

Much of East European higher education suffered from the perception that success in a few fields of fundamental sciences extended to philosophy and to the social sciences. The academic results and training in the latter fields bears little international comparison. Scholarship in fields such as sociology or political science can be assessed against relatively stable views on good scholarship by particular international research communities, the aim of the East European social sciences has been much narrower. The primary aim of the state-socialist philosophy and social sciences was to justify the existence of a particular ideology and political regime. It was largely unscientific and lost its relevance with the collapse of the political regimes it lauded. Using the liberal quality discourse, one might argue it was of reasonable quality since it met the purposes set by the former political regimes and that it lost its quality because of rapid changes in external conditions. One may, however, wish to see education upholding qualities more universal than those defined by a particular political ideology, no less in the West than in the East.

While quality of scholarship can be assessed within the relatively stable context of international research communities, higher education is more than purely academic training. Higher education has at least two further elements, considerably more context-dependent than basic scholarly knowledge and research skills: personality development and the development of certain transferable skills.

It is often said that traditional British higher education has much to do with the molding of *gentlemen*. State-socialist higher education had its own personality-forming mission. This was to mold a *"tovarishtch"* or "comrade": a person brainwashed with a particular ideology, hating capitalism and private gain, ready to sacrifice their life for the global revolution. It is not surprising that the East European regimes used higher education extensively for forming the communist personality. So long as state-socialist higher education was geared toward the development of heavy industry, political loyalty was considered to be a necessary, if not the most important, element of any professional qualification, for example an engineer.

The enthusiasm that many of those trained by the old higher education system devoted to the overthrow of the regimes they had been trained for and to restoring much that was contrary to their training, says something about the success of the personality-forming program. Its remnants are probably disappearing from universities. What they are replaced with is not fully clear. Perhaps it is an ideal of a rational profit-seeker that newly emerging market economies expect higher education to produce. However, as nation building is considered to be an important task in many of the newly independent countries, one can also expect universities to put a serious weight on the development of national consciousness. In the 1990s, the rector of a leading Estonian university published his thoughts about the university's role in nation building, views that would have been quite appropriate for the Central Europe context of some 60 years earlier.[28] The second half of the 1990s suggests that the recovery of national pride is an important task for Russian higher education.[29]

While state-socialist higher education stressed memorizing large quantities of facts and the soundness of the communist soul, transferable skills are gradually coming to the attention of higher education. Knowledge of foreign languages and computer skills in particular are more attractive for private businesses than rigorous academic knowledge. The availability of information technology is one of the issues scrutinized during the accreditation process in most of these countries. Still, few students have regular access to computers. Foreign-language faculty was the first to leave universities for better remuneration in the business world. While there is a tendency to list the general skills helping a graduate throughout their life career, these skills are also context-dependent: even information technology develops so fast that the computer skills, if not specified further, cease to be useful. That many East European graduates have built successful careers abroad indicates that the system had many unintended outcomes developing, for example critical thinking—not only critical about capitalism as it was supposed to be, but also itself. Reassessing the essential purpose of higher education in different contexts is a major issue not only for bureaucracies and business administrators.

Conclusion

Higher education is a heterogeneous field, so attempts to assess it against a single universal standard are unlikely to succeed. Certainly, a strong desire exists in the East European countries to confirm that the quality of its higher education meets international standards or is catching up.[30] It is driven largely by particular political programs. The reason why these countries are trying to find vaguely defined universal standards instead of more substantial local ones is related to the threat of a far-reaching public debate occurring on the purpose of higher education for the previous state-socialist political regimes. Such a debate may all too easily lead to the conclusion that a significant part of higher education lost its *raison d'être* as the old political and economic establishment collapsed.

Yet, current quality research in the West has something to learn from recent East European history. The state-socialist countries tried to relate higher education to the perceived needs of society to an extent unprecedented in the history of the university. Defining the quality of higher education as infinitely contextual, to the extent that any factual evidence that questioned particular ideology was declared in error, corrupted higher education and particularly the fields of social studies and philosophy. Profitability as a criterion of truth could well have a similar impact on higher education.

Higher education is contextual to a significant extent, particularly in those aspects related to the development of particular skills and the cultivation of certain personal characteristics. With diversification and the move from strictly academic to vocational training, higher education becomes increasingly related to local needs rather than the requirements of global research communities. This may, however, be offset by principles of academic integrity and certain universal values. Newly established East European quality assurance mechanisms are driven by many concerns including internal and external politics, interests of particular universities and academic groups, as well as by the need to secure social stability. Yet, its connection to education remains relatively weak. In the long term this may be a serious problem. In striving for greater homogeneity, the Western world may soon be faced with a similar situation.

Chapter 9

Reproduction of the "State Nobility" in Eastern Europe: Past Patterns and New Practices[1]

> Who are our enemies? Who are our friends? This is a question of the first importance for the revolution.
>
> Chairman Mao Tse-tung, *Little Red Book*[2]

Throughout the history of mankind, the way nobilities tend to reproduce themselves by passing wealth and status along family lines has been an issue of conflict between those who have capital to pass on and those who have none, over the history of mankind. Revolutions offer rare opportunities to radically change the relationship and overthrow what some revolutionary thinkers have called *reactionary classes*. When this happens, the chain of reproduction is broken, often leading to a situation where the best capital one can possibly trade with is having no capital whatsoever. Not having economic or cultural capital may then be a necessary, if not sufficient, condition not only for access to power, but may also become critical in physical survival at a time when Cultural Revolution sweeps over a country. This is, however, a rare occurrence. Soon, the newly established nobility needs to legitimize itself once more, through means other than brute force.

In modern times, power legitimizes itself through educational systems. In his study, Bourdieu describes how the French *state nobility* reproduces itself through the system of the *grandes écoles*, demonstrating that the argument about democratization of education through its massification remains, to a large extent, a myth.[3] Access to educational tracks leading to positions in the nobility remain as closed as they were decades, if not centuries ago.

Since its very beginning, the Soviet Union tried to follow a radically different path: to develop a truly democratic and massive educational system and prevent accumulation of cultural, or for this matter, any type of capital that could be converted into power. The results were mixed. In its time, the Soviet Union developed an educational system of unprecedented coverage. Related to this, various systematic measures were taken to facilitate social mobility. While moving youth from collective farms and industry to "institutes," as the majority of higher education institutions were then referred to, was a relatively easy task for the totalitarian political regime. The opposite was also to be facilitated. People with higher education qualifications were to be constantly rotated away or their impact neutralized by other means. To learn the lessons of the past, the difficult question to be addressed is about the point from which measures, supposedly leading toward more democracy, have an opposite effect. The recent East European experience may offer a good reason to rethink some old issues.

Some comments are offered in this chapter concerning the production and reproduction of elites in Eastern Europe, with the primary focus on the Soviet Union and its successor countries. It seems that the main beneficiaries of the so-called revolutions of the late 1980s and early 1990s were the last generation of second-rank communist *nomenklatura* who effectively converted their former status into economic capital and very rapidly learned the ways of reproduction that Bourdieu so well describes.

A word should be said about the usage of *Eastern Europe* in the current context. It is not a geographical notion as it reaches further into the east from anything one could possibly define as Europe; neither is it a political term, as it covers countries with NATO and OECD membership. Here Eastern Europe is understood as the European *other*, what Neumann calls the *barbarian at the gate*.[4] If nothing else, then the FBI teaching law enforcement in Hungary, for example, supports this latter comment. *Eastern* also includes a part of the Asian *other*. Despite the statements made by politicians in all the relevant countries about their deeply Western hearts, the author here has taken the liberty of defining the whole former state-socialist camp, with the exception of the former East Germany, as the *other*. The fact that the author, by virtue of writing this chapter, has become the *other* to himself should, at least to some extent, help to avoid the reading of this text as yet another attempt to divide the world into good capitalists and bad communists or vice versa.

Bourdieu on Reproduction of the Elite in France

French higher education and its role in what Bourdieu describes as reproduction of the national elite remains in many respects unique. While the universities seem to have little say about decorating the rulers of the country and its businesses with necessary attributes, a group of institutions usually considered to be representative of the lower ranks of prestige and referred to as *vocational higher education* dominate the field of elite reproduction. However, one might agree with the argument that one particular relatively closed group of schools holding a monopoly of elite reproduction is probably more important than the way the schools are historically named, or the substance of the education they provide. The latter, as argued elsewhere,[5] is becoming increasingly irrelevant.

Following Bourdieu, it is not enough to study one or another elite school to understand the nature of elite reproduction in any given country. Changes in the prestige of certain fields study or in admission procedures may often leave an impression of democratization, however, studying a whole group of closely related institutions one would soon find that positions surrendered by some representatives of the group are immediately occupied by other institutions, often more elitist and with even more restricted access. This is one of the main arguments of Bourdieu's study—the massification of higher education, as occurred over recent decades in France and other countries, does not necessarily represent the democratization of higher education. While a closed group of institutions are responsible for converting economic and cultural capital into symbols of power, massive numbers of university degree-holders do not have access to higher positions in business or in the state apparatus. As Bourdieu puts it, the symbolic value of a diploma is proportional to its rarity. It seems to be true that the increase in the total number of degree-holders inflates the symbolic value of mass-produced degrees. Highly valued symbols are produced in the sectors that remain relatively closed: "It was necessary to bury the myth of the 'school as liberating force,' guarantor of the triumph of 'achievement' over 'ascription,' of what is conquered over what is received, of works over birth, of merit and talent over heredity and nepotism, in order to perceive the educational institution in the true light of its social uses, that is, as one of the foundations of domination and of the legitimation of domination."[6] This pattern is by no means unique for France: "In 1984, a mere 13 élite boarding schools were found to have educated 10 percent of the board members of large US companies and nearly one-fifth of directors of two major firms, as compounding exclusive college degrees with upper-class pedigree multiplies the probability of joining the 'inner circle' of corporate power."[7]

Drucker states that the reason for this is that: "Developed countries, especially democracies, are convinced that they need a ruling élite. Without it, society and politics disintegrate—as, in turn, does democracy."[8] One might wish to add that it is not only democracies that are afraid of disintegrating in the absence of a strong elite.

Playing an important role are transformations with different types of capital and elite schools as the locations of exchange—rather than as the temples of knowledge that some would prefer to see them.[9] Access to higher positions in the state machinery and corporate leadership, especially that of French companies, is as a rule facilitated by the grandes écoles. However, the field of grandes écoles itself is diversified. Bourdieu divides the schools between the *grande porte* composed of the top grandes écoles where often 60 percent or more of the students represent the dominant class (schools like *École Nationale d'Administration, École des Hautes Études Commerciales, Institut d'Études Politiques de Paris*) and the *petite porte*, which have a relatively small proportion of upper-class students, often 35 percent or less, engineering schools, humanities, and science faculty. According to this division, the grande porte leads to "most noble careers in industry, business, higher civil service, and research, and... facilitates future transfers from one sector to another."[10] While the petite porte leads to executive positions, jobs as technicians, mid-level managers, secondary school teachers or, at best, narrowly specialized "minor engineers, jobs that rarely permit career or transitions from one sector or job to another."[11] As with some other religious doctrines, Bourdieu's sociology shows that the more common path, the petite porte, despite all its worldly attraction and ease of pursuit, does not lead to true salvation.

The gate, however, is not everything. Where the door leads to also depends on who is knocking upon it. As Bourdieu demonstrates, the level to which the top grandes écoles can facilitate upward social mobility is limited. Statistically speaking, there is always a certain, if small, proportion of students from the dominated classes in the top schools. However, despite the most highly valued credentials they do not enter "the most noble career paths," trying to rely on merit without having the necessary social and symbolic capital for them they find their employment as secondary-school teachers, middle-level managers, and so on. Thus the grande porte becomes a petite porte. The opposite also seems to be true—representatives of the dominant class, especially those with the "*de*" prefix in their names, or those whose families have held high positions for several generations, can enter the most noble career path through the petite porte.

Finally, it is important to note that in the French case, the legitimacy of the top schools in producing highly valued symbols comes from the traditionally highly centralized and powerful state. Under the pressure of internationalization of the political sphere and the pressure of the markets—particularly as represented by multinational corporations—their legitimacy has begun to erode. As a reaction to this, some of the traditional grande écoles are trying to reestablish themselves as international, let us say, business schools. For example the 200-year-old school of bridges and roads, *École Nationale des Ponts et Chaussées*, is widely advertised as an "international business school."

Formation of the Elite Under Communism

The function of elite reproduction can be performed within different institutional settings, as Levy's study on Latin American higher education indicates; depending on the history and structure of the higher education system, an elite can be produced both by private as well as public institutions.[12] The nobility can graduate from the nonuniversity sector or research universities. The main requirement, however, seems to be relatively limited access and admission from selected social strata.

At least in European tradition, the state has provided the university with legitimacy and protection to carry out its reproductive mission. Amongst other things, until recently, the state had the ability to protect the university against the markets.[13] Alongside the massification of higher education this is becoming history, the state cannot afford large numbers of people enjoying academic leisure at its expense. But it must secure those who give it its own legitimacy because it is not only university that needs the protective hand of the state. The opposite is also true, power needs a trustworthy institution standing above everyday mundane concerns to furnish it with symbols of higher wisdom, the MBA degree is a somewhat extreme but increasingly popular example. This cements the foundations of a symbiotic relationship between power and the elite-producing sector of what was once higher education, and what some Britons have recently started to call *the knowledge industry*—mutually taking care of each other's needs for legitimacy, a process that may or may not be related to knowledge.

Even groups as hostile toward intellectuals as the Bolsheviks in communist Russia were, and later in the entity called the Soviet Union, had to turn to the higher education establishment for legitimacy despite all their distrust and hesitation. Certain changes had to be made and old values had to be reconsidered to enable higher education to better fulfil its new task. Lenin made his position on the matter very clear, "The intellectual forces of the workers and peasants are growing and getting stronger in their fight to overthrow the bourgeoisie and their accomplices, the educated classes, the lackeys of capital, who consider themselves the brains of the nation. In fact they are not its brain but its shit."[14] Communist ideology had no need for university with its idealistic mission of searching for the truth because the final truth had already been revealed to the Party, making a large part of traditional intellectual work suddenly redundant. There was a very practical need to pass the message to the *progressive classes*, workers, and peasants, for this literacy was to be promoted, which offers an explanation of the educational success of the Soviet system. The content of the revelation, however, could not be challenged. Berlin explains: "The new directive was that the business of the intelligentsia—writers, artists, academics, and so forth was not to interpret, argue, analyze, still less develop or apply in new spheres, the principles of Marxism, but to simplify them, adopt an agreed interpretation of their meaning, and then repeat and disseminate and hammer home in every available medium and all possible occasions the selfsame set of approved truth... The celebrated Marxist formula—the unity of theory and practice—was simplified to mean a set of quotations to justify officially enunciated policies."[15]

Thinking as a creative process became redundant. Agreement with the doctrine was no more a matter of intellectual debate but an issue of mental health as "Dissidents have been declared insane on the grounds that any healthy person would recognize the virtues of the Soviet system."[16] The role of Marxist theory in East European communism or rather *state-socialism*[17] remains unclear. Some say that one should not blame Marx for what was done in communist Eastern Europe, maintaining a view according to which the "Eastern bloc... is seen as a misuse, a perversion of the Marxist theory...."[18] Others, including Soviet reform communists of the 1980s, tried to combine the maxim "More Socialism More Democracy"[19] not only with rescuing Marx, but also Lenin. Still, others like Todd adopted a radically different position, maintaining that "Marxism–Leninism does not explain such economic perversions as the Soviet Union. It causes them."[20] As long as one has not seen the successful practical application of Marxist theory, Marx's own approach would not allow us to support him. This may well be one of the reasons why many of his former followers have distanced themselves from him recently. But, as history has not yet come to an end, one may leave this issue unresolved for the moment.

There were particular reasons for the communist regimes' tolerance of certain types of education, and what is more, subject to close supervision of the representatives of the progressive classes, often students of working-class origin, the "lackeys of capital," were allowed to participate in it. This was only possible after the radical reform of the educational system, and the abolition of its symbols and academic ranks in particular, separating professional fields of study from universities and subjecting them directly to the requirements of industry, and opening admission to higher education to virtually anybody, but with a preference for representatives of the progressive classes. The communist regime had no need of philosophers to nag at its *philosophy*, but it did need teachers to preach the philosophy and teach the illiterate masses to read so that they could read it. Preparing for the world revolution (preparations continued until the end of communist regimes in Europe and most probably continue in the remaining strongholds elsewhere) the regime also needed scientists, engineers, and qualified workers to build up industry and develop weaponry. This was the education, mostly in engineering institutes, that became responsible for the formation of the communist elite, which Chomsky calls simply "the military-bureaucratic élite."[21]

The first generation of communist leaders established their capital, however, in political prisons, battles, and intrigues. While Lenin studied for three months at the

Kazan University, Stalin's cultural capital was less abundant. "He (Stalin) was a half literate member of an oppressed minority, filled with resentment against superior persons and intellectuals of all kinds, but particularly against those articulate and argumentative socialists whose dialectical skill in the realm of theory must have humiliated him often before the revolution and after it..."[22] providing just another excuse to exterminate those asking too many questions. Working-class consciousness was considered sufficient to decide even upon academic debates. Stalin was, among other things, glorified as the leading linguist. Leonid Brezhnev, holder of an engineering degree from a third-rank institute, became, in time, a leading national intellectual and was awarded, among other things, the highest award in the country, the Lenin Prize, for his literary works. The picture was, however, more complex than that. For example Gustav Naan, the president of the Academy of Science of the Estonian Soviet Socialist Republic over three decades from Stalin until Gorbachev, and who never finished his undergraduate degree in physics in Leningrad despite his appointment as a loyal party member, expressed "viewpoints that have been recognized as valuable from the scientific point of view."[23] This stands in stark contrast with the reassessment of his activities during the *perestroika*, which almost led to his expulsion from the Academy. The story goes, that the only reason why this did not occur was an attempt to avoid a repetition of the precedent of Albert Einstein in Nazi Germany.

Soon, higher education became once again a necessary step on the ladder to nobility. There seem to have been three necessary elements a combination of which paved the way to the highest ranks of bureaucracy or a career in industry: party, school, and higher education at an institute. The relationship of upward-moving nobles with education was not always exactly that of ordinary students. In 1929, Khrushchev moved from the Ukrainian coal mines to Moscow at the age of 35 and was admitted to study at the Stalin Industrial Academy. He, however, became the secretary of the Party Committee at the Academy, that is the highest official within the Academy. In 1931, he moved further in his party career, becoming the party secretary for Moscow in 1935 and, as a matter of fact, the Mayor of Moscow in an era of extensive execution of counterrevolutionaries. The path of Nikolay Ryzhkov, prime minister under Gorbachev, was not very different: from a mine to heavy industry in the Urals, meanwhile joining the Communist Party, graduating from S.M. Kirov Urals Polytechnic Institute, moving to highest levels of management of the Uralmash plant (the same institute and plant from which Yeltsin came) and moving to Moscow in 1975.

While the pattern seems to be explicit, there are some significant deviations. Edward Shevardnadze, foreign minister under Gorbachev, and later president of the newly independent Georgia, originates from a teacher's family, passed through a career in the Communist Youth and Party organizations, but apparently has little industrial experience. Finally, Gorbachev himself, a world-renowned intellectual superhero, accumulated his initial capital at a collective farm that paved his way, however, not to an institute but the Moscow State University, from which he graduated with a law degree. Gorbachev's case may also to some extent contradict the conclusion of a recent study by Mokhov, according to which the elites originating in agriculture are the most conservative.[24] At the same time, his long career clearly follows the common pattern among his group that Mokhov describes as the "long lift," decades of patient serving and gradual promotions through various offices.

Exchange of Capital

The pattern emerging here is not utterly different from what Bourdieu describes in his study. First, the communist elite did not come from the university, but from the vocational sector, meaning that countries are run by people who, in addition to the party school, had often been trained as technocrats. Second, close ties are established between the state and business. However, if in the French case careers seem to progress from the state to the business sector, the East European nobility moved from industry to the highest levels in the state bureaucracy. One major difference is, however, that the communist regimes were very sensitive about accumulating, or actually not accumulating cultural and symbolic capital.

There are significant differences between the patterns of capital exchange and accumulation that pave a person's way into the nobility. The pattern Bourdieu develops is linear—symbolic and cultural capital leads to a proper educational institution. Matched with the symbols of a proper education, this capital lays the foundations for a successful career with swift movement between the state bureaucracy and private business. As mentioned earlier, it is not so much the sum of capital that matters, but a proper mix of various elements. In the East European communist tradition accumulation of capital along family lines was avoided, sometimes by extreme means. Two methods were used to facilitate the systematic destruction of cultural capital under the communist dictatorship—political terror and affirmative action. Political terror, meaning extensive execution of those who posed a threat to the Party leadership through their cultural and symbolic capital, was exercised more than once in recent East European history. Lenin applied it in the aftermath of the revolution, Stalin continued the practice in the 1930s during the campaign known as the "*Yezhovshchina*," executing more than a million people, largely Communist Party members and including the most prominent. Stalin initiated another massive purge in January 1953. The first event in the new campaign, known as the Doctors' Plot, was aimed directly against intellectuals and was anti-Semite in nature. It was, however, interrupted by Stalin's death in March 1953. Bloodsheds in Hungary in 1956 and Prague in 1968 can be interpreted in terms of a restoration of the same policy in the Soviet-controlled territories of Central Europe, giving the *younger brothers* a lesson in the methodology of proletariat dictatorship.

Another way to avoid the accumulation of cultural capital along family lines was by opening access to higher education to students from collective farm and industrial backgrounds. Successful implementation of these affirmative action policies earned high praise internationally, although, to remain fair in the face of history, the international organizations who played these tunes were often funded by one particular member state. Still, the educational system through which a poor peasant youth could obtain a law degree and become president, or indeed virtually anything, attracted more than enough attention from Africa, Asia, and Latin America.

The champions of democratic education yesterday and today are campaigning to relate access to higher education to academic merit alone, implying that eliminating economic and social status-related criteria from among the access conditions for higher education would promote social justice. The issue here is that merit at entry to higher education is too often predetermined by status or, as Giddens puts it: "In any case, a fully meritocratic society is not only unrealizable; it is a self-contradictory idea... In such a social order, the privileged are bound to be able to confer advantages on their children—thus destroying meritocracy."[25] The only way to promote truly democratic higher education is to radically separate it from performance-related criteria, a policy occasionally tried out in the Soviet Union and elsewhere, for example in Italy.[26] On top of all the positive characteristics, similar policy may have raised one major question concerning the extent to which an institution called upon to carry out this policy should be called educational.

The policy of affirmative action was introduced by Lenin in 1918, his decree opened higher education admission to everybody of 16 years of age or above, not depending on previous learning experience. Similar approaches have been tried since in other democratic countries. One recalls, for example the Swedish 24 + 4 scheme. Every person of 24 years of age or above with at least four years of employment experience is entitled to enter higher education.[27] Chuprunov et al. describe similar practices from the latter days of the Soviet Union: "In addition, it is a widespread practice in the Soviet Union to allow admission to VUZ without examination for specially advanced workers chosen by their enterprises for higher studies. The selection of such candidates takes place during workers' meetings, after examination of the proposed students' files."[28] The authors go on to explain, "this practice is considered an encouragement for the high quality of their practical work."

Considering the somewhat narrower mission of communist higher education as already mentioned, particularly its stress on memorizing dogma, the damage caused by the admission of academically less qualified but loyal students might not have represented such a major problem, that is without mentioning the meaning of *academic performance* in the given context. From the historic perspective, one has to admit that this policy, if applied for long enough, backfires on the economy, as well as on the culture. It does seem to be the case that destruction not only characterizes the Soviet sphere of cultural exchange, Gaddy and Ickes suggest that the whole Russian economic system fails to add any value to its natural resources, but rather destroys it.[29] Why then would one expect the educational system to work in the opposite direction and create value?

Capital has a strange tendency to accumulate, meaning that children of higher education graduates tend to have a better chance of entering and succeeding in higher education in France, Eastern Europe, and other places. At times, the communist leadership took the initiative in overcoming this by establishing stronger ties between higher education and industry, a familiar issue these days in British higher education, for example. The massive reform initiated by Khrushchev in 1958 required the students entering the institutes to work full-time during the two initial study years, attending the institutes in the evenings.[30] This was to confirm, among other things, their working-class loyalty. This campaign, like many others, failed relatively quickly and what remained was the privileged admission of working-class students. This meant a requirement to have a document certifying employment experience, something that the children of managerial staff acquired relatively easily, concealing the hidden accumulation of cultural, symbolic, and social capital.

Giddens confirms the creativity of the privileged in finding ways around the official policies: "...even in the relatively egalitarian soviet-style societies where wealth couldn't secure the advancement of children, privileged groups were able to transmit advantages to their offspring."[31] Matthews' study, however, establishes a different kind of connection between the Soviet families' economic status and educational opportunity of their children. His argument is that the parents of poor families lack the necessary connections and are unable to bribe those who need to be bribed to secure the admission of their children.[32]

The official policy promoted by the communists in Eastern Europe was very much that of democratic education in its extreme form, turning higher education into a stay at an institute as a special reward for industrial work (e.g. Khrushchev) or work in a collective farm (e.g. Gorbachev) and exceptional political loyalty. Looking at the Soviet affirmative action policy in the context of political suppression, one may, however, conclude that this says less about the democratization of education than it does about policy on preventing the emergence of elites apart from those engineered by the political establishment.

The right mix of capital that led to the highest positions in the bureaucracy combined a degree from the mass higher education sector, party school training, a progressive career in heavy industry, unrecognizable personality and exceptional political loyalty. The process, however, had inherent conflicts. The pattern Mokhov presents is that of a tension between conservatives driven by the "long lift" to the highest positions and the rapidly rising young technocrats, "novices," to be rotated to the political and geographical peripheries almost as soon as they reached the positions of "seconds" in the various power hierarchies.[33] Roper describes a similar policy applied in Romania: "...Ceausesçu instituted a policy of élite rotation that he used within the party, government and military. Although the principle was never formalized, élites were periodically reassigned to new state or party positions, including rotation between the national and the county level. This policy ensured that no individual could consolidate power within the party or at the local level."[34] For many, a political career within the state-socialist system has been what the game theory would refer to as a "one shot game": someone gains a position, grabs as much as he (or extremely rarely, she) can, and prays that he or she can retire in peace.

The New Field of Elite Production

According to a widespread myth, the spell of communism in Europe was broken about a decade ago, leading democratically elected peoples' representatives to power. At least, this is what the new leaders from Bratislava to Vladivostok claim happened during the revolutions, a corrupt elite has been replaced by enlightened free-marketeers who apparently emerged from who-knows-where. Interestingly enough, not very many of them were unknown before. They were often the holders of second-ranking positions who had been living under the threat of being marginalized too early by top-ranking conservatives.

The first step in the formation of the new elite and establishing a new elite production system is privatization. The process goes something like this. On day 0−1, before the privatization, everybody has little money. On day 0 privatization starts. On day 0 + 1, some people privatize almost everything, and most of the people privatize nothing. Some call it the token economy, where the value of a token depends on who has it.

Through privatization, positions in the bureaucracy and the communist power hierarchy have been translated into wealth. The story of the young Russian *oligarch* Vladimir Potanin is a good example of post-communist transformation of capital. Being the son of a Foreign Office employee he entered one of the most exclusive schools, the Institute of International Relations in Moscow and received an excellent education in foreign languages, gained fluent French, good English, and so on, and became responsible for procurements at the Ministry. This job enabled him to establish his own company in the early 1990s; channeling some of the Ministry's functions through his private firm he became within a few years, one of the richest and most powerful men in the country. Now, he owns a bank, three newspapers, mining, oil, and telecommunication businesses and, surprisingly enough, has the reputation of an honest businessman. This last point has caused major problems for the Western investors who invested in that particular trait.

This is by no means unique to Russia, as many, particularly in Central Europe—which remains a part of the *other*—love to argue. Frydman and colleagues have exposed the hidden side of post-communist privatization of state assets:

> According to a former leading politician in Rumania, 80% of new Rumanian millionaires were part of the Ceausescu-era nomenklatura; many had been in the arms industry and have since built their fortunes on arms trading (skirting the arms embargo on Croatia and Serbia was a particular boon). A survey on Russia's top one hundred businessmen complied by Moscow's Applied Politics Institute found that 61% of the country's new rich were members of the ex-nomenklatura. A Polish economist who traced the careers of several hundred nomenklatura from 1988 to 1993 found that over half of them turned up as to private sector executives. The numbers in Hungary are reported to be even higher than in Poland.[35]

As these authors argue, through the process described in terms of the transition "from *nomenklatura* to *kleptoklatura*" or "from plan to clan"—the old *nomenklatura* gaining title to state assets received more power than they had had under the communist system, which had often proved to be rather insecure. Playing on suppressed nationalistic feelings is often crucial in securing a power base, particularly in the former Soviet Union, but also elsewhere: "The East Central European state-building with roots in dissident movements already possessed nationalist inclinations—Vaclav Havel and Lech Walesa come to mind—but virtually all the post-communists, who had traditionally rejected 'bourgeois' appeals to the nation, accepted the national discourse in order to avoid marginalization and compete effectively in the political arena." Even more striking was the behavior of state-builders in the former Soviet Republics. Almost without exception, republican communists adopted the language of the nation in order to forge alliances with national popular fronts before independence, and to maintain their political base among key elites after independence.[36]

Having arisen from the revolution, the next thing the new elite had to do was to cement its status with exclusive educational symbols. Finding a noble alternative for the proletariat-related engineering institute is a crucial issue here. A representative of the Ukrainian Ministry of Education indicated the direction that the demand of the *new elite* for *new degrees* has given to higher education systems, declaring, "there is a need to reduce the surplus of engineers, and increase the number of lawyers, economists, sociologists and managers."[37]

Recent structural changes in East European higher education are largely related to the establishment of the training field for the new elite. More than a thousand new higher education institutions have been established on private initiative over the past decade in the countries of Central–Eastern Europe and the former Soviet Union—at least 340 in the Russian Federation alone, hundreds of new higher education institutions have been established in newly independent states of the Transcaucasus; at one point, the number of private universities in Romania also approached 100. They are, as a rule, all oriented to fee-paying students and define their mission in terms of promoting an education that meets *Western standards*. Two fields of study seem to prevail in these schools, business and law, the fields of the new elite.

Traditional higher education institutions are also fighting for a monopoly over the mission of elite-formation. As Savchuk et al. report, almost all 200 public higher education institutions in Ukraine have established private branches to offer fee-based education in business and law.[38] For new private institutions, as well as for traditional public universities, business and law are the fields that attract the students able to meet an annual fee of $2,000–4,000 when average monthly incomes are around $100. It is a degree in law or business that the newly rich, old nobility thinks will guarantee their offspring a prosperous future.

In an era of innovation in higher education, reproduction of the elite does not end with traditional higher education institutions. In his study Darvas describes the Invisible College, Budapest, an explicitly elitist complementary tutorial program designed to "re-create the academic traditions embodied in institutions such as the British college system, the French École Normale Superieur, as well as Hungarian predecessors such as the 100-year-old Eötvös College... IC tries to ensure that the tutors focus equally on general skills which members of the future élite are expected to possess, including communication and self presentation."[39] Many consider the program highly successful. It has already produced Hungary's premier. It has also been exported to at least eight countries, including Kyrgyzstan, where it is called, without any hypocritical modesty, *The School of Future Elite*. While a large number of new and traditional universities are trying to monopolize the field of elite reproduction, many of them are no more than mere diploma mills delivering promises that cannot be fulfilled.

As the field of elite training is highly competitive, the issue of legitimacy arises. During the period of political turbulence, the state lost at least part of its power to protect higher education institutions, as well as the money needed to do that. The old higher education's legitimacy in producing the new leaders has also been challenged. It would not be a big problem if there were not so many new service providers trying to monopolize the lucrative business of elite production. Following Bourdieu, "The prince is only able to get truly effective symbolic service out of his painters, his poets, or his jurists insofar as he gives them the capacity to legislate within their domain."[40] The painters, poets, and jurists may choose to undermine the poor prince and look for other sources of legitimacy confirming the *fundamental law of the economy of legitimation*—the law of the proportional relationship between the symbolic efficacy of a legitimizing discourse, as Bourdieu puts it, and the visible distance between those praising and those praised. This has moved the focus of search for legitimacy beyond national borders. Masses of new universities call themselves "European" and "International," implying that the symbols they produce have stronger guarantors than the local feudal versions. While this quite often proves to be an outright lie (there is no European or international agency that could confirm the status anyway, and the questions remain open as to what education in economics means for the Russian virtual economy or education in law for a lawless country like Belarus or the standards that should be applied), accreditation of East European higher education institutions is becoming a lucrative business for many Western agencies, as well as for the market-driven higher education institutions. Failing to recognize Western accreditation could be political suicide for an East European bureaucrat—being employed by a new democratic country—one that should be open to new thinking.

The state on its own part also takes steps to support its legitimacy, often referring to international agencies. A representative of the Albanian Ministry of Education clearly points to his agency's fount of legitimacy: "The main aspects of this strategy are determined in new legislation based on progressive European standards supported by, and implemented with the help of, the Council of Europe."[41] One of the most remarkable initiatives belongs to that of the Romanian legislator who reacted to the challenging post-Ceausescu growth in the number of higher education providers with a requirement that all new study programs be approved by Parliament. A recent report compiled by the Council of Europe indicates that all the post-communist countries covered address the need to develop quality assurance mechanisms usually as defined in terms of state accreditation,[42] which indicates the existence of significant tensions within the national systems. How strange these procedures are that they are usually justified with recommendations from Western experts serving one or another international agency. Foreign agencies have been given a substantial role in reestablishing the post-communist field of elite production where masses of institutions aim at the establishment of the monopoly and its inscription in stone. It is, however, unfortunate that service fees are playing a much more important role here than a sense of responsibility.

Conclusions

Communist regimes in the Soviet Union and elsewhere in Central and East Europe once tried to establish an elite production system that was not based on accumulation and exchange of various forms of capital—cultural, symbolic, and economic—just the opposite, they related it to the destruction of that capital. As the Bolsheviks had tried to outlaw money from their economic system, they also tried to destroy cultural and symbolic capital, initially by executing the holders of that capital (killing the nobility, as well as the bourgeoisie and its "accomplices," and later Communist Party members were executed if they came to accumulate capital from other sources) and by considering loyalty to the official doctrine as one of the most important admission requirements to higher education.

While the ultimate program of the communists is clearly about physical and symbolic destruction, the minimum agenda adopted since the late 1950s required constant confirmation of the prospective elite's identity with the historic mission of the proletariat through factory work, various memberships, and the party education network. The minimum program left the bureaucracy some margin to corrupt itself and convert social capital and the little money it made into privileged status for its offspring. Furthermore, it violated the dominant code of ethical behavior, and the law and the society became more corrupt, reaching its peak with the massive privatization of state assets in the early 1990s.

The universal corruption of communist society established the scene for the peaceful revolutions of the late 1980s and early 1990s through which the former second echelon—pragmatic enough with its technocratic background and nationalistic passions—seized power and translated its social capital into economic wealth through the successful privatization of state property. To further legitimize its status as the new nobility, it now requires a new field of elite education to contrast the former system of mass higher education in engineering, its own *alma mater*. More than enough providers of exclusive training in business and law have emerged recently to satisfy the need of the aspirants for the new nobility for distinct symbols. One of the special characteristics of the new elite training field is that it clearly separates itself from mass education, primarily through charging substantial tuition fees.

The legitimacy crisis is a characteristic of the new state, as well as its higher education. Competition for the production monopoly in elite symbols erodes the power of the state even further. A new fount of legitimacy is found in various international or foreign agencies that often see, in legitimizing the new power, a particularly lucrative business, often paid for by the European Commission or other funders. It is still to be seen whether these processes will stop after a new equilibrium is established, or whether they will continue to further erode the power base of nation-states. It is, however, discouraging to see that the capital used in exchanges is in many ways surrogate, based on the social capital that accumulated in the offices of local party leaders and legitimized through diploma mills. Bourdieu's message is more positive—it is always like this. There are no saints, at least among those in power. Humboldt expressed the view that many have these days on the new nobleman of Eastern Europe, "we may admire what he does, but we despise what it is."[43]

Chapter 10

The Communication Community and the Scam of the Knowledge Society[1]

International conferences, symposia, workshops, seminars, and the like take place literally every day on higher education. The World Bank, UNESCO, the OECD, and many regional organizations, as well as universities have all recently organized meetings to discuss topics such as higher education in the twenty-first century. Such gatherings are usually carried out in a spirit of progress and optimism. For the past 30 years or so, higher education systems all over the world have grown rapidly and nowadays many view higher education as a source of national prosperity just as much in the East as in the West, North, and South. Unfortunately, what policy makers, consultants, academics, and other wishful thinkers ignore is the hard fact that the present in which the rich nations exist has only a tenuous connection with the future of the poor. It seems that higher education is only making the rich richer while the investment of poor nations in higher education brings meager returns. One could argue that they should try just a little harder, or that an increase in higher education funding of another 0.5 of a percentage point of the GDP would trigger a quantum leap toward prosperity. Unfortunately, there is no empirical evidence for such a causal connection. There is some bitter truth to arguments like Foster's, to the effect that educational expansion is merely *the cart not the horse of economic growth*.[2]

Although many see higher education as a source of wealth, there are not too many wealthy universities around. Quite often the university has become a place where those who can afford it, go to spend a few years in a quandary trying to decide what they really want to do with their lives. Public funds to pay for such parking places are becoming increasingly scarce. There are many impoverished universities, and not only in the traditionally poor regions of the planet. In Europe, albeit Eastern Europe, there are universities that lack desks and chairs although they still claim to be preparing *national elites*, and that is without mentioning the lack of information technology and the other great inventions that would enable future leaders to prepare for the global utopia to come. In many places, the economics of higher education has been turned upside down. While discussions on the importance of higher education for economic development continue, it is frequently the case that the next statement delivered by a cabinet minister who has acknowledged the crucial role of higher education in economic development concerns universities taking care of their needs by themselves, perhaps casting some doubt over the sincerity of praise made earlier. There are many good reasons why a significant number of young people should be kept, for a certain period, from entering already overcrowded labor markets.

Particularly beneficial, at least in economic terms, would be doing this at the students' own expense. However, one should not forget the proportion of students whose studies are by no means related to the development of, or even contribute to, any intellectual activity. With the swelling cohorts entering postsecondary education, it may soon be the case that a higher education qualification will become a precondition for applying for almost any kind of paid employment.[3] From this it follows that the content of higher education is being reduced to the level of *transferable personal skills*. These are particularly fashionable in countries like the United Kingdom and involve communication, teamwork, and so on, skills that one might well, however, expect young people normally to have developed by their mid-teens.

While none can doubt that the most successful economies relate to impressive higher education systems, not all massive higher education systems relate to economic success. Some are related to complete disasters—economic as well as political. In its day, the Soviet Union developed a higher education system of unprecedented scale simply to demonstrate to the Western world the success of its ideology and meet its military industry's need for engineers. Many learned from the Soviet experience and established *open, free,* and *high quality* higher education. To disseminate the Soviet experience on planned educational development, the UNESCO opened the International Institute for Educational Planning in the 1960s, which still operates in Paris. Sometime later, however, it became strangely clear that no economic success related to that massive educational system. It also became evident that much of the humanities and social sciences in that higher education system dealt primarily with the dissemination of rather unsophisticated ideological justifications for the support of a particularly violent and antiintellectual political regime.

Under such pressures the initial debate about the great achievements of Soviet higher education was modified to demonstrate that the fruits of its success were consumed by the military industrial sector that developed nuclear weaponry, the space program, and so on. Recently, this claim has also been brought into question by reports suggesting that up to 70 percent of Soviet military technological developments represent the success of the most brutal campaign of industrial and military espionage that human history has ever witnessed. Evidence of the successes of state-socialist higher education grows ever more tenuous. Recently, others have argued that Cuba's large higher education sector may fall into a similar category.[4] One may only wonder with all the evidence available on the recent collapse of the state-socialist bubble, how feasible it is for educational systems worldwide to open up to ever-growing numbers of students on the assumption that somehow the relevant content will appear from somewhere. It is no great surprise that the wallets of international consultants are filled with convertible banknotes, local elites—often furnished with symbols of wisdom from oversees universities—become corrupt. What is missing is development. Whatever the rationale behind the OECD educational indicators, one should remember that indicators themselves produce nothing, and for political reasons national governments quite often inflate their indicators—as the USSR once did—just to look good. But so long as we fail to clarify whom we train and for what purpose, 5 percent enrolment may do as much good for the economy as 20 or 30.

However, the intention of this chapter is not to open yet another debate on the economics of higher education. Nor is it about public and private returns from investments in higher education—many lifetimes have already been spent on that task. One may feel a certain commiseration for the men and women of integrity who have exchanged their academic positions for offices in international financial institutions and who iron out the wrinkles from the shirt of global capitalism. Still, it is a choice they made for themselves. This chapter is not even about higher education as a motor for economic development. One can easily collect hundreds if not thousands of such papers printed over the past decade. Some would argue that this indicates the rapid growth of knowledge. Others, and I would count myself among them, would maintain that the total volume of printed paper does not necessarily correlate with the amount of knowledge. Rather this chapter discusses a few issues in higher education that are not necessarily measured in dollars, bahts, rands, and so on. It is about the place of higher education in the peripheries of the capitalist economy, culture, and other aspects of higher education which, by the end of our era on this planet, may matter more than growth and development quantified by the number of cars, computers, rockets, and nuclear bombs we have been able to consume for our own benefit and to the disadvantage of many others. Perhaps, it is time to start thinking about how mankind could do more about what really matters for itself, uniquely, as humans with limited resources. One reason for this might be that by spreading the globalizing, beer-drinking Western student culture to other parts of the world, or by making developing countries pay for a random selection of Western trivialities delivered through global distance learning consortia, is a waste— if not simply immoral. Still, the chance of introducing intellectual relevance to higher education, which is increasingly becoming a marketplace for global *providers of educational services*, remains slim. Although educational systems, many of which have their roots in training civil servants for colonial regimes, do not carry locally relevant intellectual missions either.

During an earlier visit to Thailand in 1997, the dean of the Social Science Faculty of one of the Thai universities told me a story that went something like this: in our university we do a lot of research on prostitution. We have a great many researchers studying this problem in Thailand. They produce a large number of research reports. However, the problem is that this does not seem to be helping us tackle the issue. Actually, just the opposite seems to be the case. As the pile of research reports grows so does the problem of prostitution in Thailand. This tale is difficult to forget. It is a story, told sincerely by an intellectually committed social scientist, that shows something somewhere is fundamentally wrong with social sciences, even though one can easily produce a philosophy of science that utterly ignores any request for practicality in the science academic. After all, what is the practical use of Big Bang cosmology anyway? A cynical response to this story came from a well-established sociologist in Moscow who said, "The purpose of teaching sociology is to train professors of sociology. That is it. Sociology as a discipline does not have any further purpose than its own reproduction."

The starting point of this chapter is rather different. It agrees with Schutz, who once stated: "The world of nature, as explored by natural scientists, does not 'mean' anything to the molecules, atoms, and electrons therein. The observational field of the social scientist, however, namely the social reality, has a specific meaning and relevance structure for the human beings living, acting and thinking therein."[5] A voice from Africa delivered the same message; Makgoba proposed, "knowledge is a human construction that by definition has a human purpose. Knowledge cannot be sterile or neutral in its conception, formulation and development. Humans are not generally known for their neutrality or sterility. The generation and development of knowledge is thus contextual in nature."[6] Such an approach falls perhaps on the periphery of internationally organized academic *tribes and territories*.[7,8] Theirs is a difficult position, caught between the demands of international research communities and local societies. How might this be changed under the pressures of massification of higher education and the globalization of what is increasingly referred to as the knowledge industry? Ultimately, however, the main issue is one of social justice—why should the taxpayer pay for an individual's personal pleasure or career aspirations if the only outcome is paper wasted on printing research reports that nobody but the authors and their close friends ever intend to read? Why should he pay for degrees that may stand for absolutely anything, including nothing.

The Economics of Higher Education for Everybody

From the economic development perspective, there are two main reasons for supporting education, including higher education. First, a more highly educated population is able to produce more sophisticated goods and consequently earn more money for themselves and thus pay taxes to their governments. Second, a highly educated population is socialized into more demanding and selective patterns of consumption of clothing, food, and entertainment. They need ever-new products that other highly trained people like themselves constantly invent—DVDs instead of books, CNN news instead of the local radio, and nuclear and biological weaponry instead of bow, sword, and rifle. This is termed "progress." It cannot be sustained for long[9] and is threatening to become its own opposite. For every good end-product there has to be demand. If there is no demand then demand has to be created—by advertisements showing celebrities consuming new brands of beer or toothpaste, or constructing images in the media of the enemies of the nation, to be killed using the latest technology.

In this scheme of things, university degree-holders have a well-defined place as producers of expensive and sophisticated goods on the one hand and then as their consumers on the other. Here, the motivation for people to learn is equally simple: in a socially mobile industrial society those with more skills and knowledge move to higher positions and are able to enjoy the most sophisticated pleasures material life can offer. Higher education consequently has a dual purpose. On the one hand, it is a part of the productive force. On the other hand, it also socializes a class of sophisticated consumers. The perspective of moving upward on the social ladder, toward higher levels of consumption, is the principal reason for the growing demand for access to higher education. Focusing on what looks increasingly like the failure of higher education to drive economic development, two possible causes can be tagged. First, higher education fails to contribute to the nations' production capacity.

University graduates, like the millions of graduates once in the Soviet Union, are not, despite their formal qualification, able to design and lead the production of goods and services that anybody would desire. Society is not able to use their qualifications. Second, it may well be that graduates from mass higher education are not that much more discerning in their taste than other classes. As we suggested earlier, massifying of higher education may not necessarily expand the social base of the *nobility*. There is significant evidence that higher education graduates are increasingly moving to jobs the nature of which does not require such education.[10] One may, perhaps, argue that while graduates have often moved thanks to their university studies from a working-class background to the middle strata, this move is neither reflected in the content of their jobs nor in the pattern of their cultural (and other) consumption. Still, the creation of additional market capacity is perhaps a significant reason for international financial institutions to support the expansion of higher education. It may, however, be the case that higher education itself is an item of consumption with no other ends beyond it, and that its capacity and mechanism for driving economic development is similar to that of a new detergent or toothpaste. Seen from this angle, the market capacity created in developing countries is that of educational services for Western, increasingly profit-seeking universities that have one single agenda—to sell.

Thankfully, real life is more complex. Patterns of production and consumption are asymmetrical. The system includes all kinds of violations such as nonproducing consumers, consuming nonproducers, and other free riders—philosophy professors for example. There are people, although fewer in number, who still resist the basic idea of modern society, which is to overcome anxiety and insecurity in this world by accumulating for the sake of accumulation alone.[11] It may also be the case that not everybody is as anxious as the modern Western man and consequently, despite everything, some people find their security beyond the realm of the material. One can imagine that there are still a few primitive peoples who have deliberately chosen not to measure their personal value in terms of the destruction they produce on a daily basis.

If one asks what kind of contribution social science research and training make in an ideal productive and socially infinitely fluid society, things look rather simple. A significant part is related to creating consumption. Market research is a traditional field for social science graduates. Another important aspect relates to the creation of highly sophisticated theories that are printed in unbelievably expensive professional journals and books and shared in conferences, the consumption of which is affordable only to a privileged few. The production and consumption patterns on that market (international social science research and training markets) are clearly asymmetrical, with clear divides between the North and South, West and East. While the developing world is forced to consume products from the developed world, the benefits may actually be less than drinking a coke or a soda everyday. Such benefits circulate as symbols, degrees, titles, memberships, and so on within the closed world of any research community. They do not necessarily relate to anything external to it.

Higher Education and Social Mobility

The first major problem one faces in considering education as an economic category is that no society is infinitely socially mobile. While only cynics say that by the end of the day everybody will be promoted to their level of incompetence, since the

Second World War only Japan has made any serious effort to reach that far. The truth is that, at any given moment, a significant number of people occupy positions for which, from the point of view of acquired skills and knowledge, better candidates are available. Likewise, there are well-qualified people who occupy positions that require considerably lower levels of qualification. From the economic point of view, this is a waste. Nevertheless, life is not only about economics. Society cannot be infinitely fluid, if for no other reason that the cost of moving everybody around all the time would be stupendous. Some thinkers, such as Morris and Linda Tannehill propose a universal solution to this problem: remove the agent responsible for absorbing the risks and shocks caused by the market—the government: "...men who had spent their lives as tax collectors for the Internal Revenue Service or as Federal narcotics agents would find no demand for their 'services' and would have to change careers in order to survive—perhaps even to that of garbage collector or janitor (honourable work, for a change)."[12]

As Bourdieu showed there are many variables in addition to formal qualification that determine one's position in the hierarchy of the consumers of material benefits.[13] Much depends on social background. Industrial society has not been able to overcome the rigid social structures of agricultural society where, given a stable technology, the rulers' qualifications did not matter too much.[14]

If from the perspective of achieving the highest labor productivity, unrestricted social mobility may appear attractive, facilitating it might be technically too difficult and too costly. In attempts to standardize education according to knowledge and skills acquired, it is sometimes overlooked that education, at least good education, is more than just a laundry list. To a high degree job qualification relates to organizational culture and contacts, the acquisition of which takes time, quite apart from the social cost of insecurity related to the permanent firing and hiring of personnel. Social mobility may in turn be limited by the plain cost of the process so long as society is supposed to take care of a proportion of those who for some reason are not productive—something the Tannehills disagree with. Yet, with the markets taking over the social sector, society *is* becoming more mobile. Massive introduction of information technology is supposed, amongst other things, radically to decontextualize the work place. If, for example an employee fails to display the right amount of political correctness, somebody from the waiting list will be hired in his place. The obdurate is then recycled in the system of lifelong learning after which they return to the waiting list. At the very end of this road, which some call free-market democracy, lies a not very encouraging picture combining libertarian markets and legalized euthanasia for those who choose not to participate.

Industrial society, and postindustrial society probably even more so, is moving toward greater job decontextualization. One can move from one position to another based on what one does rather than on whom one is, where what one says matters more that one's credentials. However, the world is not there yet, and it is not obvious whether cultural and social capital can be destroyed to such an extent that their influence can be ignored. Legitimating is still a big business and will probably grow rather than decline with the growth of the number of degree-holders on the labor markets. While big businesses develop their own training programs, in many things trusting that the university (whilst keeping an eye on its performance) is more efficient than training or testing every employee. When half the population holds a university degree the university one graduates from and field of study becomes even more important than in the situation when 5 percent, the crème de la crème, graduates and almost automatically occupies the noblest positions in society. There are many worthless universities, and their identity is rarely a secret.

Bourdieu put forward the idea that the effectiveness of legitimation is proportional to the distance between the legitimating agency and the legitimized.[15] To put it simplistically, it is more effective when somebody important from a distant country says that I am a good guy than if I do it myself or ask my brother to say it. The leaders of many developing countries have traditionally been trained abroad. Not always because the content of the training they received is more relevant, often the opposite is the case, but because it provides them with credentials that have higher degree of legitimacy and more highly valued than those from the neighboring university.

Relevance and Legitimacy of Higher Education Studies

The basic rule of legitimation contradicts a commonsense understanding that higher education, knowledge, and skills, as well as the development of personality characteristics should be relevant to the graduates' employment in any given environment. This presupposes, of course, that universities are not expected to provide absolutely universal education. Increasingly, degrees designed according to the needs of academic *centers*, a limited number of economically developed countries, are automatically legitimate across the entire *peripheral* world. That studies abroad offer a fertile ground for "brain drain" suggests something is going slightly awry. More often than not graduates choose not to return to their countries of origin, for primarily economic reasons. Graduates are looking for the best living and working conditions they can possibly find. It is certainly the case that hardworking students from Asia or Africa are extremely competitive among the spoiled children of welfare societies whose primary interest is in extending their childhood. However, while the economic motives are certainly present, there are others as well. First, a cultural divide discourages graduates. Having been socialized into Western culture it is difficult to return to an environment that, as has often been explicitly stated, is of inferior quality. Narciso Matos, one time secretary general of the Association of African Universities lists seven reasons for highly trained people to leave Africa, among them is the following: "long graduate and postgraduate programmes in developed countries, [are] often in areas unrelated to developmental needs, or [are] not coupled with strategies for reintegrating the trainees in their home countries and communities of origin."[16]

By and large Western educational systems, which are increasingly aggressive in attracting students from overseas, pay little attention to the elements of education that would not only help students to return, but also facilitate change within their home environment. The logic of Western higher education and perhaps the entire European enlightenment project is that of one-dimensional progress. According to this, the United States or Germany look today like Nigeria or Bangladesh will look, if not tomorrow then at least the day after tomorrow. This false vision supposedly justifies the training of future leaders for developing countries according to programs designed for the needs of the developed economies. Second, despite all attempts to homogenize cultures, peoples, and places through globalization—the trivializing power of global capitalism—much of the contribution that social sciences are expected to make remains largely context-specific. Thus, students returning from Western universities should be able to identify and solve problems as they are perceived by local societies, but this is not something they are trained to do. The gap between the two positions may be significant. It is not a rare case for a donor country to first contribute a problem and then come up with a solution. The recent spread of a Western novelty called *gender studies* into Eastern Europe perhaps serves as one such example.

Separation of mind from body, message from medium, value from fact, and form from content has been a fundamental principle of Western philosophical thought since Aristotle. Almost automatically it calls for the development of universally applicable methods—problem-solving strategies that would apply anywhere with equal success, on the presumption that the content-variables are isolated with surgical precision. Unfortunately, imposing binary divides on social realities always contains an element of arbitrariness. These days many follow Karl Popper who once declared that *Alle Leben ist Problemlösen*[17] [The whole of life is problem-solving]. They repeat the mistake of the great philosopher who failed to see that many human problems were socially constructed and thus infinitely contextual. This would perhaps explain some of the failures Western academics and consultants have experienced in different environments. It also calls for caution as one tries to calculate the value of a Western university graduate in an African village: "... high-level cognitive resources embodied in African graduates returning from overseas training consistently define and delimit solutions to problems in terms of the particular understanding of development, or economics, or education to which they have been extensively exposed while abroad."[18] It may well be the case that the only value such a graduate embodies, is by nature, symbolic.

The *Higher Education and Adult Studies Program* at the University of Denver, Colorado offers a Doctor of Philosophy degree that according to the brochure attracts students from "the Rocky Mountain Region, nationally, and around the world." The benefits of the program are explained as follows: "Having one of the largest concentrations of public agencies related to higher education in the country outside of Washington, D.C., provides students with unique opportunities to grapple with public policy issues. Moreover, the state capital, with its legislature and the National Council of State Legislatures, provides forums for understanding the state policy as it relates to higher education." It does not take an advanced degree in public policy to see that relevance is at issue here. While Bourdieu's legitimation rule may well work and a doctor of philosophy degree-holder from the University of Denver can get a reasonably high position in many places in the developing world, the extent to which understanding public policy in Colorado helps to develop higher education in a country where a university has no desks or even electricity remains somewhat of a moot point. From such a context, the individual returning to a teaching position will soon let their students know what a great place Colorado is. But they may need more. The problem is, as Vilakazi puts it, "Western-educated Africans become lost and irrelevant as intellectuals who could develop African civilization further."[19]

Globalization

Discussion about globalization began almost imperceptibly in the 1970s and has recently reached colossal dimensions. Few respected social scientists have not contributed an article if not a book to the issue. After a quarter of century of the accumulation of writings, some argue that the very concept is becoming too wide to carry any meaning. Others say that one should perhaps talk about different kinds of globalization. A third group, as always, declares the whole phenomenon to be yet another hoax. Claus Offe, for example states, " 'Globalization' might also be considered [by future historians] as something typical to our epoch—even if it is mysterious how this term, which is so overloaded with diverse meanings, could ever capture the minds of contemporaries so thoroughly."[20] To the extent that globalization as a concept has a central idea, it is that under the impact of globally organized, borderless business and information technology the world is becoming a *global village*—to use the term coined by media guru Marshall McLuhan back in 1964[21]—and that many of the differences are simply being eradicated. In its popular form, it is a somewhat naïve narrative about global, democratic, liberal, market capitalism wiping out poverty, injustice, and oppressive political regimes. Capitalism is expected to bring large-scale prosperity while populations hooked to the Internet exercise permanently direct participatory democracy, voting with their home PCs. Making a profit globally together is the goal that many expect to render traditional ethnic, cultural, and religious differences irrelevant. Again, not everybody shares that view. Some may still believe that humanity may have more important purposes than simply maximizing profits. Others say that profits are always made at somebody's cost and that the cost of the welfare of Western civilization is paid by slaves in the developing world: "Slaves keep your costs low and returns high.... Slavery is a booming business and the number of slaves is increasing. People get rich by using slaves. And when they've finished with their slaves, they just throw these people away."[22] Given what Bales learnt about sweatshops and organized slavery all over the world, one may wish to conclude that more than sharing wealth, globalization allows more intensive exploitation of poor nations by further weakening national governments. While the nations' desire for the fruits of economic and technological development is normal and natural, documents like a recent OECD report clearly indicate that our planet cannot sustain consumption comparable to that of currently industrially developed countries on a much larger scale.[23] Unless this problem is solved, differences between rich and poor nations or groups within nations will remain and even get worse.

The popularity of globalization has allowed academia to construct yet another field of study with its barely accessible jargon and a whole range of academic products, the purchase of which counts as cultural and symbolic capital. As always, as soon as a new academic tribe is legitimately established it becomes self-sustaining. There is no further need for it to relate to anything external. Obtaining the legitimate right to sell symbols of knowledge and wisdom, university degrees can sustain a discipline far beyond the exhaustion of its content. The global academic market allows the sale of a phenomenal number of academic products of extremely varied quality under the simple disguise of its *Western* origin. Those whom the new discipline ought to concern the most, the poor of the developing world, could hardly care less about an exclusive group of academics talking and writing about their poverty in a professional jargon accessible only to a privileged few.

Globalization is somewhat a strange topic for higher education anyway. University is an institution that, by its very definition and its roots in the enlightenment and *reason*, has already overcome the divisions between cultures and nations. The spread of the university globally, particularly during the second half of the twentieth century, should itself be considered as one of the early signs of globalization, paving the way for the rationality of science instead of being influenced by it. This is the modern university born in Western Europe that carries globalization to various corners of the world. To be more precise, what the university allegedly carries as globalization is faith in reason and an infinite trust in science. "Science" here means *hard science*, the united body of knowledge that one day is expected to express all the processes with a single universal formula. The ultimate science is collecting and processing data that one day, many say that very soon, will give us the ultimate *theory of everything*.

Two problems arise here. First, while traditional Ivory Tower science sees nothing wrong in spending immense resources on creating the ultimate formula, this may not help much in solving any particular problem, and even if it does help, it may still be far too late. Second, starting from the other end, solving the problems of particular people may not support the ultimate academic aim of the scholars, and may even threaten to divide the privileged high culture into problem-driven subfields, to the detriment of the global, but exclusive communication community. The fragmentation of knowledge is one of the main problems that the academia thinks it is facing. It is not so often realized that not fragmenting knowledge may have even worse effects. The problem higher education faces in moving to the various corners of the world is finding a balance between the university as a temple of reason, and local cultures and needs that may not prove to be reasonable at all, at least from the narrowly rational point of view of a Western scholar. Three possible solutions may be offered. First, to put our trust in globalization and its capacity to level out all significant differences between nations, cultures, and various localities as, for example Sztompka[24] does. Second, to deny globalization and make a major effort to develop all possible local differences to their extreme limits, for example condemning written culture as a symptom of Western cultural imperialism, and building institutions of higher learning in some parts of the world based on oral tradition and a different form of communication than academic journals and conferences—with ancestral spirits, for example. The third way, as always, would be a middle way: accepting the principle of reason as a reasonable one and trying to apply it for the benefit of various local communities. "The task we have before us, then, is understanding the situation of the contemporary University without falling into either nostalgia for national culture or the discourse of consumerism."[25]

In discussing globalization, the global nature of the university as established on reason under the modern program is not what is usually meant. Rather, the issue often evoked concerns the relationship between the two modern institutions of the university and the nation-state. Perhaps the nation-state should be blamed for imposing its own agenda on the university, often at the expense of reason which, according to Readings, should not be the case: "So on the one hand, the state must protect the University in order to ensure the rule of reason in public life. On the other hand, philosophy must protect the University from the abuse of power by the state, in limiting the rule of established interests in higher faculties."[26] Readings's position seems to be based on the assumption that science can reach its totality within a single national culture: "Science has its unity in the framework of a national language, which forms a closed totality within the wider totality of absolute knowledge. Hence science is grounded in a historical ethnicity rather than purely rational abstract idea of a people as pure will."[27] Yet the universal idea of reason should prevail over ethnicity, particularly as the idea of the Prussians as the only true and cultured nation, perhaps the only ones capable of reason, should be pondered a little more deeply. Delimiting the power of reason by national boarders and languages is a violation of idea of the university itself.

Achieving Local Relevance

Recently, critical voices have been raised against the West for imposing teaching methods and content on universities in developing countries, which fit neither the local traditions nor the environment. Universities in Africa, for example were once established by colonial regimes for the purpose of providing civil servants, not to promote local cultures and indigenous people's interests or solve their problems: "Though physically located in Africa, the colonial university did not respond to the needs of the country: this was not the true objective for which it was founded."[28] Tukumbi Lumumba-Kasongo asserts, "...the essential and common objectives of the colonial powers were to provide a modicum of education to fit Africans into the colonial expectations and systems of production and consumption."[29] The introduction of scientific research in colonial countries is similar, "... 'modern' science was introduced, just like formal education, in the wake of colonization, and its primary goal was to facilitate the economic exploitation of local resources for the benefit of European metropoles."[30]

International financial institutions like the World Bank in its recent tertiary education strategy[31] explain the need to develop higher education in transition economies by a neo-Schumpeterian argument namely, the capitalist economy being driven by technical innovations.[32] Regrettably it ignores the alternative, neo-Smithsonian view. According to the latter, the market drives the capitalist economy. Unfortunately for developing countries, Western development aid does not create as much market demand for local goods and services as it creates demand for Western products with which it enslaves those economies even further. As they fall further into the trap of foreign debt they are made to pay for the Westernization of the consumption habits of local, often corrupt elites: "...the existing developmental models, in both their political and economic reforms, have essentially served a few African political elites and their patrons and institutions in the North or the West rather than African people and their social systems."[33] The role of higher education graduates is not overly different from those directly under a colonial regime. They are still representatives of another world. Colonialism continues in a different form, which, according to the logic of capitalism, is perhaps even more efficient than the previous political colonialism.

Academic colonialism reaches far beyond university lecture halls. It is actively involved in constructing common everyday knowledge. Recently the Western world celebrated rather noisily the millennium, taking the end of the second millennium as a hard empirical fact. What the once Christian Western world seems to have forgotten is that millennium is not an established empirical fact but a socially constructed event. Many other cultures have constructed their own calendars in different ways and within their context the millennium does not relate to anything particular. Buddhists, Moslems, Jews all have different views on what year it is. Europeans in their ignorance and arrogance, however, do not see any problem with going to Tokyo or Jerusalem and celebrating the arrival of the new century there, at the same time teaching the local, and what are considered to be less advanced cultures, how the true reality appears, as if thanks to European civilization humankind has finally realized its real purpose. Nkomo, for example writes, "These formerly silenced indigenous knowledge systems should be brought back to life through a renewed African research in education process informed by the scientific method of inquiry. History is full of examples of how civilizations rise and fall. And the current dominant Western civilization is not 'the end of history.' "[34] Although, there are strong signs that European civilization may be the last before mankind commits collective suicide, it is perhaps too pessimistic or too arrogant to think that there is no alternative whatsoever.

Some may believe that centralism is to bring civilization to barbarians. What they actually do is trivialize the world through the often-violent construction of a simplistic world culture. Gellner remains somewhat sceptical about the achievability of the latter: "The emergence of a single universal culture may yet come: only the future will tell. Nevertheless, for the time being, what we see is the replacement of enormous cultural diversity by a limited number of high cultures with political pretensions. This is the age of nationalism."[35] Whether we really need it is another question. There is of course another, who apart from the operating salesmen, needs a single culture, high, middle, or low?

Still, we have not yet reached the limits. Western education, propelled by its own ignorance, has been taking over educational sectors in many regions they consider to be underdeveloped or not developed at all. This usually means education in a foreign language, and the delivery of factual knowledge and cultural values constructed from the Western perspective. It is, for example a historical fact known by any student in the Western world that Columbus discovered America in 1492. Native Americans may, however, have a minority view on this as Seepe points out: "Thus, instead of saying that Columbus 'discovered' America, native Americans might prefer the phrase Columbus 'invaded' America. And instead of saying Livingston discovered Lake Victoria... it might be closer to the truth to indicate that Livingston was led to the lake by Africans."[36] There is obviously a limit to which Western higher education can be adjusted to the needs of radically different societies. Getting the facts right, however, is something it ought to do for the sake of its own values. So far, it seems that the Western world is interested more in exploiting others and imposing its own ways as the global high-culture up on the emerging global village—or still better global marketplace—rather than trying to change its own habits, which are not always sustainable.[37]

While questions about the relevance or accuracy of particular Western-produced knowledge in other contexts can be easily raised without compromising the *reason*—just the opposite—lack of reason and the limits of knowledge in the West may well be one of the main causes for Western scholars sticking to their narrow understandings. Locally constructed methods by which knowledge is created poses a more difficult challenge however. Western science is based on certain principles that have to assure the objectivity of academic knowledge. The laws of nature—if measurements are carried on in the same, correct way—remain the same regardless of ethnic origin or religious beliefs, conscience, and so on, of a particular scientist. In this respect, scientific knowledge is by definition radically decontextualized—it is always and everywhere the same. Some, even within the Western academia, question the premise of the objectivity of scientific knowledge even in the area of natural sciences. Michael Polányi, a Hungarian-born chemist and philosopher of science, for example described how laws of nature just disappeared in his chemistry lab. The results of experiments depend on the people who carry them out.[38] Such facts have not been able to challenge the mainstream thinking of the philosophy of science.

However, if one talks about the need to localize the higher education establishment, making it relevant to local cultures and the ways in which indigenous people gather knowledge on their external environment, a particular question arises about the extent to which this can be done without violating the very principles on which the Western university has been established—for instance the replication of an experiment. Peter Gale recently explored some of the ways indigenous people in Australia gather knowledge by *dreaming*.[39] Many cultures have developed powerful oral cultures. In principle, at least a question can be raised about how far dreaming or meditation can be a way of data collection in an institution of learning, or whether higher education can be pursued without a written culture. For Gellner, one can talk about formal education only as long as knowledge is codified, that is presented and transmitted in a written form.[40] This may set strong limitations on those who propagate a radical return to local wisdom. Operating within the modern paradigm there is a limit on how far one can go in an attempt to find the roots.

The demands of those who speak on behalf of indigenous populations against European domination are far from clear. Some challenge the Western exploitation of the developing countries' resources and people, whilst remaining implicitly or explicitly within the capitalist paradigm. Mokubung Nkomo of the Human Sciences Research Council of South Africa is clearly satisfied with the success of some countries in becoming a part of the global capitalist economy: "Moreover, in the last half of the century there have emerged a group of countries that, the recent economic difficulties notwithstanding, are claiming a noteworthy slice of the global marketplace."[41] Such advocates see the resolution of current problems by universities focusing on the problems in their immediate environment instead of serving the needs of Western patrons or the interests of their own academics conducting internationally publishable, though not always locally relevant, research: "Tertiary institutions that are supported by the public treasury must engage in research that will benefit the public in ways that will elevate the general quality of life of the populace and thus increase Africa's competitiveness in the global arena. Given that the majority of Africans live in rural settings where entrenched traditions prevail, a great deal of the research should be devoted to addressing the conditions in this sector and the recommendation of appropriate policy options."[42]

Others, however, extend their claims further. Narcisco Matos, for example argued for a radical return to African ways of knowing and dissemination of knowledge: "Even today, an overwhelming majority of university students in Africa are the first of their kin to have access to a university. Most are part of the tiny minority amidst their kin who can read and write, who have had access to school. And yet, society in Africa does have its artists, healers, musicians, hunters, rainmakers, midwives, priests, fishermen, sages and others. Their school and education, except for occasional 'rites of passage,' is life in its entirety, their education takes place at all times, at all ages, in all places. Knowledge is transmitted to selected people, from master to apprentice. Mass education is not part of the local educational paradigm."[43] Herbert Vilakazi from South Africa called for nothing less than a "massive cultural revolution." His view is that African intellectuals should return to "ordinary African men and women" and receive education from them on African culture and civilization. From this knowledge a new African "high culture" should be constructed.[44] This prescription for Africa looks like a particular type of socialism, not greatly different from the example once established in the entity known as *the Soviet Union*. Perhaps, he has not noticed that similar attempts, particularly in Asia, have not solved any of their problems so far. Reducing the teaching of physics at university in the People's Republic of China during the Cultural Revolution to three basic machines or the even more radical approach of the Khmer Rouge created more problems for decades to come than they ever succeeded in solving. Now in Africa too, many demand a radical discontinuation of Western approaches to knowledge. They require a return to particular African ways of knowing: "Indigenous scholars in Africa and elsewhere are convinced that these spiritually centred wisdoms should serve as the springboard from which the current system of education can begin."[45] Seepe asserts, "... efforts in educational and economic development failed because they were not informed and reflective of the culture, experiences and aspirations of the majority. Africanization of knowledge thus refers to a process of placing the African world-view at the center of analysis."[46] Since the writer is far from being an expert on African epistemologies, it is somewhat difficult for him to say how far traditional African ways of knowing overlap with modern European ways. Literature available suggests the overlap is minimal. A radical return to African or other indigenous epistemologies would perhaps entail establishing learning institutions grounded on oral tradition, on spiritual or shamanistic ways of gaining knowledge. Such a strategy may be beyond the fundamental principles of the modern university. The European university can make use of indigenous knowledge by recording, analyzing, and selling it according to its own principles. But, it cannot accept it as a legitimate way of knowing. It would be even difficult to ascertain whether the modern university would be able to have a meaningful dialogue with an indigenous African counterpart.

Another issue that the proponents of indigenization of African and other societies and their learning institutions seem to be ignoring is that by discarding what they consider to be the evils of capitalism, they also reject its fruits. Nor is it obvious whether they are ready for such a sacrifice. If radical African leaders succeed in establishing an indigenous African civilization, it is in all likelihood not going to be driven by material production as is European civilization. However, there is no way to have one-way capitalism—the fruits without paying the full price. Clearly, a significant part of the price of current Western well-being has been paid by colonial and developing countries. And the West/North is quite interested in continuing the situation. But promoting capitalism in developing countries means only one thing: their societies will be vertically divided and certain groups will pay a price for the introduction of capitalism far heavier than the European countries have ever paid—there is no further periphery to be exploited to this end. Given the current international economic regime, it is extremely difficult to see how any significant part of that burden may be shifted back to the developed countries. The alternative to free-market capitalism is, and one has to be clear about it, that the material fruits of capitalist production will not be available to the developing world, which then cannot be called "developing" any longer. It would be a different (closed) civilization, perhaps similar in some degree to some of the closed countries that still exist. The example of the Soviet Union and other communist countries shows clearly that capitalist production and its related benefits like scientific and technological progress are impossible without the accumulation of capital. Here is the choice developing countries face: do they accept global capitalism, exploitation, and its logic driving toward trivialization and the bottom-line, or do they remain on their own? The third way is always an attractive choice. So far, however, the experience is that the second way, state socialism, is no more and the third way is also the first way—liberal free-market capitalism. Some believe that capitalism's evils may be contained by assigning a larger role for the state. This scarcely works in transition and developing countries for the single overriding reason that to maintain a large redistribution function is extremely costly. Regulating through markets is perhaps easier than controlling a large number of easily corruptible high-status government officials.

It cannot be ruled out, at least in principle, that by returning to indigenous wisdom a much better civilization can be established than the little-loved European civilization. Still, as the example of the People's Republic of Korea shows, there is a significant element of risk. Neither is it certain that a civilization established on the principles of spiritual wisdom will be able to feed its people. However, as Offe recently argued, thus far only one type of society has proven to be both necessary and sufficient to avoid mass starvation:[47] liberal democracy—an inherently Western invention.

In principle, there is no reason why a learning institution cannot be established on principles radically different from those of the modern Western university; for example collecting information in ways that in the West would be considered as a trifle extravagant. Carlos Castaneda is perhaps one of the most well-known authors in the West who describes such strategies in his numerous books. John Lilly, a marine biologist, described his conversation with representatives of civilizations located in other solar systems while in deep trance under the influence of LSD in an isolation tank. He offers a colorful picture on alternative ways of knowledge gathering.[48] This, however, brings us face-to-face with an old problem well known to Descartes: how on earth can we be sure those spirits are not cheating us? Descartes knew that the accuracy of the content delivered by lonely spirits wandering in the outer reaches of the universe or beyond could not be guaranteed. Perhaps, it is the public verification of the accuracy of information that makes the modern Western university such a success.

There is no reason to suppose that *Homo sapiens* is biologically modeled for a particular kind of learning institution, just the opposite. As neuroscientists have shown, the human brain is so flexible that learning could possibly occur within frameworks radically different from those currently dominant. Hence, there is no good reason to think that the modern university is the ultimate institution for learning and knowledge production. Other ways could prove more efficient or effective. It may be too late by the time Western civilization becomes aware of this in 30 or 40 years. If by then, accelerating consumption and waste of resources cannot be controlled by reason but continue to be driven by selling in the name of selling, buying in the name of buying, and accumulation of capital in the name of accumulation, the decision mankind has taken will prove to be misguided and it will be too late to know it. Imposing the primitive rules of the market on universities, forcing them to spread knowledge that sells rather than the truth according to our best understanding of it, is a grave step toward eroding the ability of the university to sustain its mission of enlightenment.

Despite the recently constructed highly mobilizing entrepreneurial university discourse in Europe and North America, the traditional enlightenment university, an institution of free intellectual inquiry is dissolving under the pressure of the logic of capitalist production. At the point at which graduates and research are defined as products akin to other products of the capitalist economy, free inquiry ends. Only such inquiry is possible that can be sold on the marketplace. Furthermore, capitalist production imposes on the university its logic of maximizing efficiency—leading to a separation and narrowing down of products and production processes. One of the most significant and recent shifts is the separation between knowledge production and knowledge dissemination. Increasingly, knowledge is being produced in leading universities and research centers, while a large part of the higher education sector, including in the developing countries has been granted the role of dissemination. Here is the invisible hand of the market that sets major obstacles to the fulfilment of Nkomo's vision: "The task of scholars and practitioners in Africa is to develop an epistemology that, while seeking a ground to itself in its specific African cultural habitat, will also aspire to be truly universal and scientific—universal in its inclusion of all humanity and not a fragment thereof (and therein lies the inexorable logic of the democratic impulse) and scientific in its pursuit of knowledge that will improve the human condition in toto."[49]

It is extremely difficult for African and other developing countries' higher education institutions to compete with Western universities in knowledge production. Moderate modernizers in African higher education would like to see a fair partnership with Western universities in developing and disseminating knowledge for all mankind. Western mass higher education can no longer afford that. Having, however, expanded beyond any reasonable limits and loaded down with all the additional social functions imaginable they have to earn to survive. What is happening is perhaps not very different from the World Bank strategy in certain countries where a high-profile consultant is paid approximately US$1,000 per day plus expenses, all charged against a particular country's loan while a local university professor earns

$50 monthly. This may be viewed as a critical investment in rapid future growth. So far, however, not too many examples of such achievements exist. Evidently, the main beneficiary here is the Western supplier of knowledge. Knowledge society rhetoric turns out to be a business scam, the primary aim of which is to prepare markets for Western academic goods and services, often at inflated prices, to be paid by government loans. Many academics have voiced disquiet about the strategies of Western aid agencies that mix development and Westernization for their own benefit.[50] Much of this echoes Readings's comments "For its part, the University is becoming a transnational bureaucratic corporation, either tied to transnational instances of government such as the European Union or functioning independently, by analogy with a transnational corporation."[51]

Developing countries do not have too much choice. Investing in Western-style higher education can bring some benefit. But, they should be critically aware that returns do not spring from just any kind of higher education. The lesson should be learned from developed Western societies. Expanding higher education in the area of "soft" interdisciplinary studies should either be considered as an investment in consumption or the creation of a parking space for an excess labor force. This despite a permanent shortage of qualified engineers to be solved by imports from developing countries. Pouring large resources into higher education, merely to meet certain quantitative targets fixed by donor agencies brings little return, "African countries began to invest heavily in education. Countries experienced a rapid expansion of educational facilities, increased school enrolment, curricula reforms and policies aimed at accelerating the pace of scientific and technological advancement. Yet, in spite of the high premium and emphasis placed on science and technology, African countries have little to show by way of progress."[52] It is perhaps less on account of the differences between epistemologies, as so much the incompetence, oversight, selfishness, and corruption both on the part of donors and recipients that vast resources have been wasted. It is not so much the idea of the university that is to blame for this but rather those who approach it from a narrow technocratic position and we manipulate it for short-term gain. As for the survival of indigenous cultures, the role universities can play here is limited. Even then, these cultures cannot be approached directly but only through the mediation of the Western academic tradition.

Political Relevance
Increasing local relevance of higher education has two separate dimensions: political relevance and substantive relevance. Both constitute particular deviations from the ideal university, the task of which is the search for universal truth. The issue of political relevance is particularly controversial. Ruling elites tend to use learning institutions to legitimate their own positions.[53] For as long as formal education has existed, those who lead societies have had to be equipped with the symbols of wisdom obtained from it. It is only during short revolutionary periods that the previous class of rulers and the educational institution of its production lose credibility, which occurred in East Europe in late 1980s and early 1990s. Then, those without formal qualification for their role could occupy leading positions. Recent history, however, also shows that even they need to start thinking rapidly about their own legitimacy in occupying the seats of power. Often they return to the university for the necessary

symbols. Thus, a symbiotic, mutually legitimating relationship between state and university is reestablished. Current debates on returning to traditional African epistemic strategies having a strong political dimension is likely. Yet, the nation-state, as Gellner argued, is not established through reason but often through lack of it. By taking on the role of reproducing ruling elites instead of cultivating reason, universities abandon the position of universalistic knowledge and retreat to constructing and justifying national ideologies. Again, developments in East Europe in the 1990s demonstrate how the production of national culture can very quickly assume a major place within the mission of higher education.

Throughout history, academia has had a complex relationship with political power. In psychological terms the dilemma that academics are caught in can be best described as cognitive dissonance. Many professors find no difficulty in arguing that they promote objective knowledge in its purest form while simultaneously cultivating national sciences. The latter concept involves a contradiction because as far as science is meant to be objective or quasi-objective, constructing a field of study for the promotion of an irrational idea of national exclusiveness for a group of people could be considered as a violation of basic principles of the academic profession. Inventing and supporting Aryan civilization and its racist identity by academics in Nazi Germany provides a warning example for those who ponder a return to the roots of ancient civilizations.

Still, university education, which by its very definition should have nothing to do with establishing closed communities based on questionable principles of skin color or particular racial characteristics, does have a major role in building nations. As Gellner explained, nations come into existence with the political institutionalization of a particular high culture. Higher education constitutes the very peak of the high culture and is needed therefore for the existence of a fully respectable modern nation. The university and its *national sciences* construct the discourse of the nation. This often means tracking down the famous who in one way or another relate to the particular nation. It also entails creating and sustaining a high language and producing new vocabulary through translating works from other languages, in brief, maintaining a nexus around which national high culture is produced and reproduced, and national elites are indoctrinated.

The politicians' expectations for higher education may reach beyond that. When times get rough and a nation faces an enemy that by all means looks very much like themselves, perhaps even speaking a rather similar language (as in the countries that once constituted Yugoslavia) but by all the *scientific* evidence is *very different*, the university is expected to produce evidence for the exclusivity of the nation. Indeed, sometimes the university is expected to produce evidence on racial superiority and inferiority, like the infamous study by Herrnstein and Murray[54] or class superiority as was done by Soviet social scientists within the Marxist–Leninist doctrine. With all the evidence one has, it would not be wise to deny the significant role universities play in creating nations. Unfortunately, accepting such a mission usually corrupts the university's academic integrity.

An interesting question can be posed concerning the role of the university as a reproductive agent of national cultures in the globalizing world. To have a truly global higher education means creating a global high culture and something some might like to call World Citizenship. What the implications of this would be for thousands of local cultures remains open. One can, however, be assured there will be enough applicants for the *world university* to produce an exclusive class of administrators for the United Nations and bureaucrats for the European Commission. The EU has actually had its own training institution operating in Belgium since the 1940s, which takes care of the socialization of those who fill the offices in Brussels. The number of all possible kinds of international higher education institutions already operating in many parts of the world offers strong evidence that such a scenario is not the result of wild fantasy.

One can possibly catch a glimpse of forthcoming global higher education and the high culture from which it will emerge by looking at the ways in which the market of MBA programs develops. Its supporting culture is the global business culture. The degrees of the best MBA programs are a valid currency worldwide, not dependent on any particular government decision to recognize or not to recognize them. So long as the business community recognizes a degree, it does not need any further formal recognition. From that perspective the power of the nation-state is already breached, maybe fatally. It is difficult to see how African nations in particular would be able to reverse that.

Substantive Relevance
For many politicians the political relevance of higher education is equal to substantive relevance. They expect universities to promote a particular, often nationalistic agenda, or to spread an ideology. Certainly, as a matter of principle such a task ought to contradict the universalistic ideas underlying the academic profession. Still, intellectuals tend to find the attractions of establishing closed enclaves tempting in much the same way as establishing ever-narrower disciplines and then monopolizing them: "In the new unit, the intellectuals drawn from the cultural zone which is in the process of turning itself into a 'nation,' can also monopolise all attractive positions, instead of having to compete with more numerous and well-established members of the group which had been dominant in the previous polity."[55] The one-time former state-socialist countries in East Europe afford many good examples of this process. Isolated from the international communication channels of academic, their social sciences were used by the political elites to disseminate their preferred political doctrine, dressed up as the ultimate stage in the intellectual development of mankind.

Closing channels of free communication, using the university to promote a particular ideology stands against the basic principles on which this institution was grounded. On the other hand, the principles themselves are not entirely without problems. It would be hard to imagine that social science has very much to contribute to the universal grand theory of everything. And while academic science would be happy to continue piling up data and moving from one paradigm to another, as Kuhn described,[56] some may say that the whole venture does not have much value if it fails to help solve particular social problems. How far the latter is achieved reflects the substantive relevance of higher education.

Academics are not inherently uncaring about the social ills of their respective societies. However, the way the university operates as a global communication community often determines the relatively low relevance of social sciences for particular communities. For academics, even slight recognition by the international research community counts more than the esteem of the local community that may be considered more limited. Balancing out, international recognition and local contribution, is not an easy task, though the international context may just as well be used only as a pretext to avoid the complexities of real-life social problems. The pursuit of *internationally adopted* paradigms may offer just as comfortable an excuse to find a quiet corner in a "rural research field" something that, according to Becher, characterizes "soft sciences" where the population of researchers is of relatively low density and problems relatively loosely defined.[57] Thus, one can claim to work on a narrowly defined field alone or with a few colleagues, and even publish internationally without being ever scrutinized by a real peer-review since there are no competent peers. This cannot happen in *urban* research fields of the hard sciences, which have a high density of researchers and careful scrutiny of results within the peer communities.

Recently, things have started to change. With the numbers of students in higher education growing and public demand for accountability for public funds spent in universities, strong forces are gaining momentum and they are threatening to dramatically change knowledge production. Gibbons and others have called the new mode of knowledge production *Mode 2*, which contrasts with the traditional university-based knowledge production controlled by disciplinary communities, which they call *Mode 1*.[58]

Mode 2 knowledge production has the following characteristics: (a) knowledge is produced in the context of application, (b) transdisciplinarity, (c) heterogeneity, (d) social accountability and reflexivity. Traditional university-based disciplinary research, however, deals with universalistic, context-free inquiry where the particular global research community sanctions meaning, questions, as well as methods of inquiry. Mode 2 knowledge production radically breaks out from that tradition and its social boundaries. It is carried out by temporary problem-oriented research groups, usually outside the traditional university research sector. As its purpose is problem-solving, it is by definition highly contextual. Its success is measured by the problems solved rather than peer scrutiny of papers produced, as it occurs in traditional disciplinary communities. Gibbons and his colleagues also claim that transdisciplinary research is undertaken more than interdisciplinary research. While interdisciplinary research is carried on in groups where participants represent particular disciplinary communities, transdisciplinary research deals with a different research ethos and culture that transcends disciplinary boundaries. Participants take on a new "identity" as researchers.

While the benefits of raising the relevance of research under the Mode 2 are obvious, the relationship between Mode 1 and Mode 2 remains complex. Although many university graduates enter Mode 2 research work in company laboratories or thinktanks, universities do not train them for such careers, but rather continue working within disciplinary boundaries, mainly because faculty career development takes place within disciplinary-oriented communication communities. Colbeck recently explored how disciplinary culture effectively prevents young scholars from researching and publishing on issues of practical concern. Instead, to progress, they are under pressure to work within the disciplinary high culture and to publish abstract theoretical papers in an exclusive professional jargon.[59]

The way traditional academic communities operate clearly stands aside from Mode 2 knowledge production. Given increasing pressure for accountability, universities should reconsider their ways of working, particularly in Ph.D. training. If graduates are effectively to join transdisciplinary application-oriented research, their initial training should prepare them for it. Universities with their narrowly specialized staff may not be ready for this. The question remains, of course, whether the university, given the way it has evolved over the past 200 years, could ever change to that extent. The natural response by the universities would be to create a subdiscipline with all its committees, journals, and conferences for each of the transdisciplinary tasks that arise. To some extent, this has emerged since the late 1950s. But it cannot be a viable solution in the long term. It atomizes academia to the point where communication becomes almost impossible with every researcher creating their own closed workshop in their own rural backyard. Yet, the knowledge industry faces a pressing need radically to reorganize itself. While rich countries can afford the luxury of legions of philosophers writing on the postmodern condition, countries with less resources should take more initiative in solving the many problems they already have. They cannot afford to finance research that serves as extremely sophisticated reading for the most affluent. It has to have practical value.

Conclusion

The crisis of the modern university often attracts the attention of philosophers, sociologists, and scholars from other academic fields. The discourse as it has been constructed over the past few decades focuses on a few common topics—access, growth in enrolment, and so on. A significant element is devoted, however, to political propaganda of a particular kind. In discussions about globalization, the global dimension and the related complexities are not always present either. Developing countries are usually exhorted to aim at quantitative targets established, for example by the OECD, in the hope that it will take them closer to a level of prosperity similar to that in Western liberal market economies. It is not too difficult, however, to see that this cannot be.

Developing countries face a series of difficult decisions concerning the development of higher education systems that contributes to their prosperity in the globalizing world. Many have suggested that Africa in particular should look toward a development model radically different from Western, liberal, democratic, market capitalism. Good historic and cultural reasons may justify this course. But, it is risky. Others suggest using the model of the Western university for the benefit of local people, but even that is far from unproblematic. The current ethos in the international world of academics, is increasingly shaped by the logic of capitalist production. It has little place for academic solidarity and often uses peripheral universities for its own profit-seeking purposes.

With the numbers of university students growing rapidly worldwide the social relevance of university teaching and research is becoming a burning issue. While elite higher education with its mission to select the smartest 2–5 percent of the population and train them for leadership positions could flourish in the sphere of abstract and often irrelevant theories imported from very different societies, mass higher education ought to be concerned with improving the lives of those who pay for it, usually taxpayers. Otherwise, it loses credibility. This is already happening in many highly developed countries where higher education is less about investing in economic development so much as investing in consumption—consumption of a few years of leisure for affluent youth and a different kind of leisure and academic tourism for many faculty. Small wonder then, that demands are growing that those who enjoy it should pay. The social function of higher education may also be changing. Strolling and holding the surplus labor force over several years is becoming socially more important than the content of what goes on there. The margin between unemployment benefit and the cost of keeping a student in university is a crucial issue from the policy point of view.

Despite the difficulties, over the past 200 years research has been organized in borderless research communities dealing with decontextualized research, and with the social sciences trying somehow to imitate the natural and hard sciences. With the expansion of higher education, clearly, university research should turn more to the problems of particular societies and relate research training to application.

Recent discussion on Mode 2 knowledge production suggests that the whole field of knowledge production is under strong pressure to change. However, traditional disciplinary research communities follow quite different principles from those that characterize Mode 2. What developing countries may learn from this change is surely to turn more decisively toward identifying and solving their problems using the knowledge available creatively, and not to copy the patterns of disciplinary research communities that belong more to the nineteenth than to the twenty-first century. The communication community of researchers, as Habermas calls the university,[60] is under construction. One of its purposes is to widen the circle of communication, which is not about researchers talking to their fellows, so much as communicating with societies at large and helping them to solve their problems. And of these, there are many.

Chapter 11

Transnational Capitalist Class and World Bank "Aid" for Higher Education[1]

> A philosopher is a useless dreamer
> he or she who is not useless enough
> is not worthy of this distinction...
>
> Victor Hatar[2]

The power of the nation-state is leaking, perhaps fatally so. From the womb of the Westphalian world of nation-states, a child has been born that now threatens the very existence of the mother. Increasingly powerful companies, once the flagships of nations, no longer accept their mother countries' dominant role. State regulations are perceived as unnecessary obstacles on the way to greater profits. Under growing economic pressure from multinational companies, states are being reduced to mere service organizations, and the names that once mobilized citizens to heroic deeds are becoming local brands, hardly comparable to such giants as Microsoft, Unilever, and others. While it is still an open question if any of the current corporate giants alone play a role comparable to that of the East Indian Company, which some 400 years earlier moved under the support of its own armed force, the total effect of the multinational companies on the human existence in the beginning of the third Christian millennium is hard to overestimate.

The emerging global economic order is assuming control over global politics. While a large number of bilateral, multilateral, and global political institutions continue to facilitate negotiations over conflicting national interests, their significance is on the decline. The World Trade Organization (WTO), the International Monetary Fund (IMF), the World Bank, and others are delivering a powerful message to the effect that politics will sooner or later surrender to the interests of economics. Empowered by modern information and communication technology, these institutions see their role as leveling the world to a global marketplace, where eternal happiness is pursued by making profits together. The consequences of such a change are massive and the role of its facilitators remains controversial or worse:

> Les citoyens doivent savoir que la mondalisation libérale attaque désormais les sociétés sur trois front. Central parce qu'il concerne l'humanité dans son ensemble, le premier front est celui de l'économie. Il demeure placé sous la conduite de ce qu'il faut vraiment appeler l' "axe du Mal", constitué par le Fonds monétaire international, la Banque mondiale et l'Organisation mondiale du commerce. Cet axe maléfique continue

d'imposer au monde la dictature du marché, la pre-eminence du sector prive, le culte du profit, et de provoquer, dans l'ensemble de la planète, de terrifiants dégâts: hyperfaillite frauduleuse d'Enron, crise monétaire en Turquie....[3] (Ramonet 2002)

While democracy continues functioning with mixed success, at the level of nation-states it is becoming increasingly irrelevant for the most important decisions made globally. No global democratically established political institution exists to control the transnational corporations and their agents, which instead of bringing prosperity to nations often follow their own greed and spread only corruption and misery, being accountable only to their stakeholders, to whom nothing but returns on investments count. Even institutions called to pave the way to global markets are becoming more interested in their own immediate benefits than in the sustained economic growth that they preach.

Globalization and the Rise of the Transnational Capitalist Class

For Sklair,[4] globalization carries a more precise meaning and a more negative connotation than the usual "compression in time and space" that proponents of the global village have been championing since McLuhan,[5] or the miracles promised by scientists and technologists. Sklair sees globalization in terms of an emerging New World order. He has developed the "global systems theory," in which the nation-states are no longer the most important actors. This theory distinguishes between the old international world and its politics carried on by states under the control of national elites, and the emerging global world order with its new driving force: the emerging transnational capitalist class (TCC).

Although until recently, companies have been subject to national interests expressed through national politics and legislation, in the new global world, nation-states become subject to the interests of a growing number of extremely powerful corporations operating above them: the transnational corporations (TNCs). Moreover, the economically less advanced countries (to call them "developing" would be a gross overstatement) that accommodate the majority of the world's population, being thirsty for foreign direct investments (FDI), increasingly compete with each other to please corporations by offering them more favorable investment conditions, such as low taxes, unregulated labor, and so on. In his study of a number of companies listed among the top 500 most successful in the magazine Fortune, Sklair argues that multinational corporations—companies based in a particular country with branches in other countries—are gradually transforming into TNCs—companies to which states no longer offer an identity. Defining globalization in terms of the new emerging world order, Sklair maintains: "Globalisation bears a contradictory relationship to the concept of national interest."[6] In the New World order there are no national interests. The role of the state is being reduced to that of a guardian of equal rules of the game in particular, geographical locations for the purposes of a growing number of corporations related to no particular country.

> Global system theory proposes that the most important transnational forces are the transnational corporations (TNCs), the transnational capitalist class, and the culture-ideology of consumerism.[7]

The new global capitalist system represents the interests of the transnational capitalist class and becomes operational through the institutions it owns and controls. The global capitalist system, however, does not only include the TNCs, but also "the policy planning networks of corporate experts, charitable foundations, and think-tanks."[8] People related to these organizations constitute:

> a new class... [with an] insatiable desire for private profit and eternal accumulation. This new class is the transnational capitalist class, composed of corporate executives, globalising bureaucrats and politicians, globalising professionals, and consumerist elites.[9]

By assuming control over the gains from global economic activities and setting new (though unachievable) consumption standards for the entire world, the transnational capitalist class is growing increasingly powerful. The drive to catch up in consumption with rich countries is one of the main forces that corrupts underdeveloped regions. According to Sklair:

> Members of the TCC tend to share similar lifestyles, particularly higher education (increasingly business schools) and consumption of luxury goods and services. Integral to this process are exclusive clubs and restaurants, ultra-expensive resorts in all continents, private as opposed to mass forms of travel and entertainment and, ominously, increasingly residential segregation of the very rich secured by armed guards and electronic surveillance.[10]

Though one may argue, following Weber, that in the hands of an ascetic Protestant accumulation may be unselfish by nature, it would be difficult to make the same argument for consumption patterns that can hardly expand wider than they have already. The TCC's consumption and lifestyles are encouraged by various forms of mass culture: soap operas, toothpaste and detergent commercials, Hollywood movies, and so on. Unlimited consumption is the driving force of the contemporary world and while for many, growing consumption does not necessarily carry a negative connotation, its very meaning—the destruction of limited resources at an accelerating pace—is sometimes ignored.

The TCC establishes the lifestyle of excessive consumption as an ideal global living standard. In striving toward this standard, further accumulation and "progress" are made possible. Achieving this goal requires "the establishment of a borderless global economy, the complete denationalisation of all corporate procedures and activities, and the eradication of economic nationalism."[11] The entire developing world and many other countries are becoming increasingly dependent on the mercy of large corporations. Foreign direct investments from transnational corporations, managed by global professionals, often mean bribes and consultancy fees for the bureaucrats in recipient countries' governments and miserable jobs for the rest. With the powerful tool of FDI at hand, "the transnational capitalist class can take control and exert its power even where there is no local dominating capitalist class."[12] It is therefore by no means surprising that the organizations in what Ramonet, echoing President George W. Bush's use of the term for the three countries that in his view constituted the main source of global terrorism, calls the "axis of evil"—the IMF, the WTO, and the World Bank—promote policies, including educational policies, which serve the purposes of the TNCs and the interests of the TCCs.

It is, however, not only or even primarily the global professionals who reap the benefits of the New World order. Executives, managers, and experts of various kinds constitute only the tip of the TCC iceberg, the main part of which remains hidden. Although for Sklair:

> The capitalist class is defined... as those who own and control the major means of production, distribution, and exchange through their ownership and control of money and other forms of capital.[13]

He fails to see that through pension funds and investment schemes, the middle classes of the developed industrial countries have become one of the main exploiting forces globally. An economic order where each participant expects returns of 20 or more percent per annum from retirement savings has turned the global economy into a ruthless war of everybody against everybody. One can perhaps paraphrase Clausewitz by saying that world economics is "war by other means." The fact that, for example a sociology professor does not manage his investments himself, by no means implies that he does not contribute to the misery of the poor in Africa, Asia, or his own country. Professional specialization tends to confuse the issue of moral responsibility, without which the current situation cannot be improved. When universities become knowledge industries, academics—once intellectuals—become workers and managers with the responsibility only to produce more "high-quality products."

Brooks in his largely anecdotal account of the Bourgeois Bohemian lifestyle, is perhaps correct in arguing that through the commodification of knowledge, for the first time in history the learned classes have gained access to individual benefits unimagined a few decades ago.[14] But, like Sklair, he fails to see the full picture that includes himself and his colleagues. Some of the controversies related to the participation of intellectuals as "experts" in the global knowledge economy are highlighted later in this chapter.

Developing Global Policies
Since the mid-1990s, higher education has become a focus for many international policy forums and discussions. At that time, during a relatively short period three global organizations—the World Bank, the UNESCO, and the OECD—all published their global higher education policy documents. This interest in higher education policy had several causes: growing enrolments and declining public funds in the Western world, developing countries' aspirations for intellectual independence through the development of their own higher education systems, and the need to provide a focus for East European countries' public policies, which had been in free fall after the collapse of the state-socialist regimes in the late 1980s. Most importantly, however, declining public funding had driven higher education systems to the marketplace and given rise to a new type of transnational company: globally operating knowledge industries or universities, like the one recently described in *The Financial Times*:

> In downtown Ho ChiMinh City—otherwise known as Saigon—stands a colonial era villa with Moorish windows and a turret, where a multinational oil company's

executives and later a senior Communist official once lodged. Today, the mansion houses Vietnam's hope for the future: an independent campus of Australia's Royal Melbourne Institute of Technology. It is communist Vietnam's first wholly foreign owned university, where around 400 students are studying language and technical skills to help them thrive in the global market. Knowledge in this small but idyllic satellite of Australia's "working man's college" does not come cheap. Tuition fees range from Dollars 1,200... for a 10-week, 200-hour English course to Dollars 8,460 for a software engineering degree course—not insignificant in a city where annual per capita GDP is estimated at just Dollars 1,400 a year.[15]

It was no surprise that higher education "service" was among the areas regulated by the WTO. The traditionally strong link between the nation-state and higher education has been weakened, and powerful global players emerge, undermining among other things the local relevance of the knowledge produced and disseminated. This is an interesting phenomenon in times when trust in a unified body of objective knowledge is being challenged on many fronts.[16]

The positions of the OECD and the World Bank, both of which are driven by global economic imperatives, stand rather close to each other. Higher education is seen as a service and expected to be run like a business. Although European countries represented in the OECD have been traditionally more cautious in promoting such views than the United States of America, which dominates the World Bank, given the perceived economic realities, higher education systems and institutions are expected to be rationalized and private initiatives and cost-sharing introduced as much as possible.[17] The UNESCO, as a noneconomic organization and one of the few legitimate channels for developing countries' voices, is doing its best to present a broader view and be sympathetic to the aspirations of the poorer countries:

> Special emphasis is placed by UNESCO on the issues surrounding government funding for higher education, which is conceived as a longterm investment for society rather than a burden on public finance.[18]

Funds to implement policies remain, however, in the coffers of the rich. The World Bank, particularly, with its multibillion-dollar loan program drives the developing countries' public policies, while the UNESCO experiences major difficulties meeting even its own running costs. For the World Bank:

> Universities in developing countries are not expected... to tackle fundamental and complex problems of contemporary society. They are more likely to be considered successful if they adapt well to aiding economic growth and do not weigh terribly much on public budgets.[19]

To track down the World Bank's success as an agent of economic growth is not an easy task. Its public documents cover issues of general development research and market economy propaganda, while reports containing project details and critical views are not gladly shared with outside commentators.[20] This may indicate a certain democratic crisis within the Bank itself: the details of projects that are ultimately paid for from the public funds of developing countries are not accessible to their public, or any other. It also exemplifies a certain ideology, put forward as early as

1965 by people such as Daniel Patrick Moynihan, who believed that the world could live perfectly well without politics if technocrats took care of the professionalization of reform. According to him, politics was reaching its shameful end as a new generation of well-trained technocrats had begun to steer the world toward greater prosperity after the Second World War:

> It is the fact that for two decades now, since the end of World War II, the industrial democracies of the world have been able to operate their economies steadily expanding levels of production and employment. Nothing like it has ever happened before in history.[21]

For Moynihan professionalization carries wide social and political implications, including the fact that in the context of steady progress assured by growing numbers of educated middle-class professionals, traditional political struggles are rendered meaningless:

> The day when mile-long petitions and mass rallies were required to persuade a government that a popular demand existed that things be done differently is clearly drawing to a close.[22]

Since those days, the idea of professionalizing reforms has spread widely all over the world. The public is fed with utopian visions of how science and technology overcome all social and natural limits by which mankind has been bound so far. International financial and development institutions, trusting in the one-dimensionality of human existence,[23] spread the gospel of politically neutral "growth." This is "le triomphe des experts" as observed by Henri Guaino, and the outcome is obvious:

> Entre les experts et les juges, il n'y a plus de place pour la souveraineté nationale ni pour la souveraineté populaire.[24]

For an individual this means that while she still lives, she no longer, as Heller says, has the "world":

> In order to have a world, however, one needs to become detached from the technological imagination—not to abandon it (for by abandoning it one would have to go native, and nowadays even that would not suffice), but to establish (to create) a distance from the technological imagination. It is the historical imagination which guides men and women in keeping this distance.[25]

Providing wider access to the fruits of capitalist production, which, needless to say, are many, may actually be a good thing, given that the pool of beneficiaries widens. Unfortunately this is not always the case. As an OECD report indicates, even the current level of consumption is not sustainable.[26] It may well be the case that the total volume of goods available worldwide cannot grow significantly beyond the current level. Therefore, we may not have a plus-sum game, which, as Gellner argues, is needed both for the legitimacy of national governments and for maintaining the democratic order.[27] We may not even have a zero-sum game. What may become the reality rather soon is a minus-sum game, where further accumulation of capital requires increasingly aggressive strategies, and increases pressure on weaker parties, such as the nonaffluent classes in the developing world. For somebody in the developed world to accumulate wealth, many more in other places have to lose. Even maintaining the life savings of the middle classes in the Western world means somebody else's deprivation. The developed world may be concerned about its *Risikogesellschaft*, but to many others, there are not even the chains left to lose.

To cope with criticism from developing countries, the World Bank has recently become rather active, at least in its rhetoric, in addressing the issue of social capital. This is an interesting turn of events, particularly given that the reduction of a society to its economic aspects leads to extreme individualization and destruction of traditional social networks:

> Risks and contradictions go on being socially produced; it is the duty and necessity to cope with them that are being individualised. The self-assertive capacity of individualised men and women falls, however, short, as a rule, of what genuine self-constitution would require. And yet this sad truth has been made difficult to grasp. Troubles may be similar, but they no longer seem to form a "totality greater than the sum of its parts."[28]

People who care for others within local communities, following principles other than offering "services" in exchange for cash, are charged with corruption. One can argue that while World Bank experts are trying to follow certain political fashions to reconcile the notion of social capital with the Bank mission to destroy it through the commodification of human relations as "services," developing a convincing theory for this may require a few neck-breaking tricks. Without this, an economist like Robert Solow may continue maintaining that the entire discussion does not make sense from the economic point of view:

> Just what is social capital a stock of? Any stock of capital is a cumulation of past flows of investment, with past flows of depreciation netted out. What are those past investments in social capital? How could an accountant measure them and cumulate in principle? I am not now worrying about where the numbers would come from, I am wondering about what instructions you would give a search party. Could one talk seriously about whether straight-line or double-declining balance depreciation is appropriate for social capital?[29]

Solow is perhaps right that, in an attempt to reduce social capital—a complex web of social bonds and connections between individuals—to a purely economic category, the concept loses its meaning. The problem with the World Bank approach is that not everything in society is reducible to mere economics. Otherwise, Bank economists would be able to calculate, for example how much social capital one Nokia cellular telephone produces, following its famous commercial "Nokia—connecting people." Even if they calculated that, it might still be the case that lending a government money to provide every citizen with a cellular phone would not create any social capital, but rather would have different and unexpected outcomes. Investing directly in creation of social capital will not create social capital, but only establish formal organizations that provide services similar to support available for free in societies with high social capital.[30] As social capital shrinks when driven by the expectation of profits, so does higher education. It loses its social and cultural value.

World Bank Higher Education Policy and Practice

A senior World Bank educationalist has recently expressed his dream of a future transnational university in rather poetic language:

> Imagine a university without buildings or classrooms or even a library, 10,000 miles away from its students, delivering online programmes or courses through franchise institutions overseas. Imagine a university without academic departments, without required courses or majors or grades, issuing degrees valid for only five years after graduation. Imagine a higher education system where institutions are ranked not by the quality of teaching, but by the intensity of electronic wiring and the degree of Internet connectivity. Imagine a country whose main export earnings come from the sale of higher education services.[31]

In his opening speech on October 27, 2001 at a workshop in Budapest devoted to the Bank's new tertiary education strategy, Salmi's message was even more dramatic, reading the afore-mentioned text over a photo of John Lennon in the background while the song "Imagine" played on a portable computer. It is particularly ironic that John Lennon is abused by transnational capitalists promoting the global capitalist society, while completely ignoring his message:

> Imagine no possessions, I wonder if you can
> no need to grieve of hunger, a brotherhood of man.[32]

Education, like many other areas, is of interest to the Bank because it is seen as a source of the modern magic: growth. The most recent higher education policy document of the World Bank makes its position abundantly clear:

> The contribution of tertiary education is acknowledged as vital because it exercises a direct influence on national productivity which largely determines living standards and a country's ability to compete and participate fully in the globalization process.[33]

According to its own words, the Bank promotes reform through its loans to national educational systems. Reform here means the increase of the public sector's effectiveness, reduction of costs and imposition of private sector logic. Later it will be argued, however, that neoliberal ideology is only a part of the problem related to World Bank loan programs. The other part is that at least some of its programs serve primarily the interests of the global professionals working directly for the Bank or contracted by it as experts and consultants. It often remains questionable whether the activities funded from the loans make any difference beyond corrupting elites in the recipient countries, while what the taxpayers receive is the growing debt burden.

In many instances the higher education strategy of the World Bank refers to the needs of the poorest countries of the world:

> More than 2.3 billion people—53 percent of the developing world—live in the 79 countries with per capita incomes less than US$885... half of them in Sub-Saharan Africa.... Of this number, 70 percent are in the 42 countries where annual per capita incomes average less than US$400. This set of very poor countries will be particularly

hard-pressed to attain provide an acceptable standard of tertiary education, even given the modest estimated cost of US$1,000 per student per year for such a standard.[34]

While the Bank seems to be worried about how to help the poorest countries to afford to pay $1,000 per student per annum, it does not see any problem in the fact that having a Bank expert working on such a country's project for a single day costs its taxpayers the same $1,000, at least as much as a local person earns during the whole year (in the poorest countries, in two or three years). There is no free lunch, and while the cost of the experts is charged against the loan, the so-called beneficiaries never see the menu.

While the Bank in its strategic documents declares economic growth as the primary aim of its loan projects, success seems to be more an exception seen in some South East Asian countries rather than a rule.[35] But even there causality seems to have been confused with correlation. Funding development is a complex issue and it is not obvious that World Bank loans delivered together with budgets negotiated behind closed doors make efficient public sector investments. It may also be the case that linear development models suit countries that have reached a certain level of development and a certain volume of modern economy. For other countries, a loan is too expensive and its use too inefficient to facilitate the two immediate tasks: to bulldoze the existing social and economic structures, and to build new ones. Perhaps the former can be achieved but not the latter. Consequently, Bank loans have positive effects less frequently than usually expected, and in order to be able to use loan money for development, a country should already be developed in the first place.

> To argue that developing countries need market-friendly policies, stable macroeconomic environments, strong investments in human capital, an independent judiciary, open and transparent capital markets, and equity-based corporate structures with attention to modern shareholders' values is to say that you will be developed when you are developed.[36]

To borrow funds to improve an otherwise miserable fixture, a country should be able to demonstrate that it has in place structures and procedures characteristic to developed countries, which is usually not the case. The result is not, however, that loans are not released; as the total amount of money lent annually is the primary indicator of the Bank's performance and it is in its interest to lend as much as possible,[37] money goes on projects whose management integrity has not been secured and that have a heavy element of spending on Western/Northern goods and services.

While it is not easy for the Bank to admit its failures, it also demonstrates difficulties in understanding the causes of problems that tend to occur in one country after another. Even if a failure is recognized, it is usually somebody else's fault, as the following quote indicates:

> In Hungary and Senegal the Bank was instrumental in supporting extensive vision development and national consultation efforts, and a loan accompanied the reform. But in neither country was the momentum of consultation fully sustained, and some of reform measures were abandoned or even reversed after political changes occurred and new actors with a different agenda arrived. These two cases illustrate also another important political economy lesson: it is difficult to promote all changes simultaneously when introducing deep reforms.[38]

Obviously, the Bank blames its failures on local politicians, unable or unwilling to see that by imposing its own policy agenda on national governments it violates some of the fundamental values of democracy that it claims to be standing for. For a bank, even for the World Bank, democracy, reform, and freedom are only words in the dictionary as long as the borrowing government commits to the technical terms of a loan.

Although the role of the World Bank as an agent of neoliberal globalization is often criticized, the criticism is usually not well substantiated because the relevant technical documentation, labeled "For official use only," is not available to the public. The World Bank strategy on higher education will be illustrated later with a few examples from staff appraisal reports of projects implemented in the Eastern European region. One of the most striking features of all these projects is the high proportion of spending on "Western" services—primarily consultancy offered by global professionals. While in some regions of the world such an approach could be justified by citing the lack of competent local technical staff, this is certainly not the case in Eastern Europe. Here it becomes obvious that more than anything such projects serve the interests of the rising TCC.

The World Bank in Eastern European Higher Education

More precise information on the World Bank's failure to implement far-reaching reforms in Senegalese higher education was not available to the author. Concerning the Hungarian case, however, multiple accounts are available for gaining an understanding on what did go awry, including one published journal article and interviews with members of the local project management team as well as foreign consultants. Following these reports, one can argue that the roots of the failure reach much deeper than an arbitrary shift in government policy after the 1998 general elections. The principal issue was that, while the Bank sought an agreement from the government, the university community of faculty and staff was not consulted for their views on a project with unexpected consequences for them.

Although the Bank had burned its fingers more than once before in Eastern Europe, lessons like the following one had not been learned:

> The lessons which are particularly relevant...are that in the field of higher education reform and management of higher education, incentives are to be preferred to compulsion.[39]

The World Bank had earlier sobering experiences also in Hungary. A World Bank internal report gives the following account of a project that was successfully completed for stakeholders and the public, Human Resources Project No. 3313 FY 91:

> That project launched a large number of new educational programs, including specific curricula and changed teaching methods. However, much funding was fragmented and lost in the morass of an unreformed higher education system. In addition, despite well-crafted objectives and criteria and a basically sound evaluation system, allocation of funds was subject to political influence and the continued spreading of funds equally across institutions, resulting in small funding amounts and limited impact.

An additional problem in this and other similar projects was insufficient protection for preserving the integrity of the peer-review process for evaluating funding proposals. Guidelines relating to conflict of interest, for instance, were not developed or monitored, nor evaluation procedures designed to discourage wholesale reallocation of funding recommendations, with the result that peer review had too little impact on decision making.[40]

This reveals major problems, if not corruption in the project management, but it does not concern only local managers in Hungary. Looking at it in the context of how the World Bank operates its loan projects, one can easily understand that given all the waste already happening, Hungarian universities wanted to do exactly what the Bank employees, consultants, and other service-providers were already doing: walk away with the little left for them after everybody else had received their pieces of the cake. One of the early missions to prepare for the afore-mentioned failed higher education reform loan project in Hungary (unsurprisingly, contracted from another TNC—Coopers & Lybrand) defined the goal of the project in rather simple terms, again ignoring earlier lessons while intervening in the most sensitive area of higher education management in any Eastern European country—funding:

> The principal aim is to help improve the efficiency of the sector and to reduce the demands it makes on public expenditure.[41]

This project was designed so that US$150 million was to be allocated to Hungarian higher education, with a significant proportion for infrastructure development, in exchange for reductions in the operating costs of the system: merging universities, rationalizing staff/student ratios, introduction of student tuition fees, and so on.

It would not perhaps be in good taste to mention money in the context of the academic profession, but the futility of expecting faculty incentives for cost reduction when professors in Hungary earn the monthly equivalent of $200–$300 should be obvious. Further reduction comes only through compulsion, and that provokes resistance. So it proved in Hungary in 1997 and 1998.

While the university community did not share the agenda of the World Bank to rationalize the structure of the higher education system and reduce the burden it posed to the public budget, the total amount of funding ($150 million) was attractive enough to avoid the outright rejection of the project by the universities. It should be also noticed that the risks and benefits rising from the project were asymmetrically distributed: while universities would benefit as organizations from the investments, individuals would carry the risks of being laid off or being exposed to more intensive workloads with no additional benefits. Such a detail apparently made little difference to the World Bank and the national government, but was critically important for the individuals whose livelihood was put at risk. So the controversy was created: on the one hand universities agreed with the conditions imposed upon them, but on the other hand worked against them using other, indirect channels, including political ones, in order to reduce the perceived negative consequences. It is needless to add that seeing consultants staying for weeks and months in the most expensive hotels in Budapest and reaping consultancy fees outrageous in comparison with local faculty pay-scales created additional tensions around the project.

The result was that in order to gain the support of the higher education community, students and their families, the opposition party's political forces promised as part of their electoral campaign to revise the increasingly controversial World Bank loan project. This was done after the opposition won the 1998 general elections. The actual process of closing the project was longer and more painful than a simple reversal. No government can say "no" to the Bank directly. Projects die slowly in great agonies even if all the parties involved know that the situation cannot be saved. For some years to come, consultants kept traveling and funds were wasted on futile negotiations. Only in private discussions can consultants express positions like the following remark by a leading British entrepreneurial university administrator brought in after the 1998 political changes: "You cannot help a university which does not want to be helped." An article coauthored by one of the leading experts on the project in 1996–97 also supports the argument that there was a fundamental clash between the agenda of the World Bank and the interests of the local academe.[42]

One of the most crucial problems for the World Bank is its lack of responsibility. The Bank runs no risk. The risks related to failed and corrupt projects, whoever is responsible for the negative results, are still carried by the borrowing countries' taxpayers, who usually have no access to the information on how the funds have been spent. As Hancock shows, not to pay its debts to the World Bank is not an option available to a developing country,[43] even if a project fails because of the incompetence of the highly paid Bank consultants. This in turn leads to the development of projects that do not present the most efficient use of public funds. Details on the spending of funds in recent higher education–related projects from Eastern Europe are presented here as an example of this.

(1) The total cost of the Romanian Higher Education Reform Project launched in 1994 was US$73.5 million. Out of that, $50 million was covered by a loan from the World Bank, with the rest provided by the Romanian government. Out of the total project cost, only $39.1 million was to be spent in Romania. The rest went to foreign experts, trainers, and others.[44]

(2) A project in Romania—Reform of Higher Education and Research cost $84 million in total. Out of that, $50 million was covered by the World Bank loan, the Romanian government provided $24.4 million and the outstanding $9.6 million was covered by a grant from the European Union. Of the total, as much as $51.4 million was to be spent abroad.[45]

(3) A project the World Bank launched in the Russian Federation in 1997, the Education Innovation Project, had a total funding of $96.6 million. Of that, $71 million was to be covered by a Bank loan, the rest being covered by the Russian government. This project had a relatively small element of foreign spending. As this project included a large Russian-language textbook to be published locally, only approximately half the money lent by the World Bank—$33.14 million—was to be spent abroad.[46]

However, it is not only the structure of spending that encourages recipients to cut corners and corrupts the project management. If one looks at what the money is spent on, a certain pattern of colonialism and exploitation emerges. A Romanian project mentioned earlier offers a pattern that is common to World Bank projects. According to the budget, one month of work by a foreign expert costs $15,000, while similar services provided by a local person costs $400. In a similar vein, training abroad cost $6,000 per month, while a month of training in Romania was budgeted at $160.[47] Approximately, one month of work by a foreign consultant was valued at the level of GNP produced by ten Romanians during a whole year.

Further, quite often there is no question about the professional competence of the consultants that the World Bank and many of the international and bilateral aid agencies use in various regions of the world: competence that leads to lessons not being learned, as in the Hungarian Human Resources project mentioned earlier. The competence of transnational consultants who work yesterday in Senegal, today in Hungary, and tomorrow in Indonesia is, to put it mildly, limited. Not having access to local languages and information sources in their expensive report-writing work, these people are critically dependent on the low-paid locals who collect and process the information and from whose reports the foreigners compile the final products for the Bank.

Eventually, one may argue that it does not matter what kind of policy the Bank promotes, as it has been promoting during the history of its existence all possible policies, while what grows is poverty. The World Bank loan projects are not about getting something done—although in certain cases some good can be accomplished—but about redistributing the tax money collected from the recipient countries' taxpayers to the World Bank bureaucrats, international consultants, and companies whose goods and services the Bank contracts. As Hancock has argued, many such companies are utterly uncompetitive in the marketplace and can survive only thanks to the aid agencies, whom they charge on average 30 percent more for their goods and services than the marketplace would allow.[48]

Instead of learning the lessons from its mixed-success projects, the Bank has learned something else: how to cut more effectively through annoying provincial political debates, and make the agreements with governments more secure in the face of possible political changes.

> Until the beginning of the 1990s, very little attention was paid to the political economy [*sic*] aspects of tertiary education reforms, on the assumption that it was technically sound reform program and agreement with top government officials were all that was needed for a change to succeed. But when it came to actual implementation, political reality often proved stronger than technocratic vision.[49]

There is no hint that there can be a problem with the technocratic vision, but only with tricking the borrowing government to buy it. It is perhaps more a rule than an exception that long-term development goals do not drive such projects, but the immediate gain of local, often corrupt elites and various global service-providers. A "positive example" provided by the Bank gives an indication of the benefits local elites may receive from accepting a loan:

> The Government of Eritrea, for example, contracted the U.K. Open University in 1998 to educate senior civil servants in a Masters of Business Administration.[50]

The least evil interpretation of such an initiative would be that for the World Bank, Eritrea is but another business enterprise, and to run it efficiently senior civil servants should learn magic phrases like "competitive edge." More regularly, however, senior civil servants, trained at a disproportionally high cost to the local economy, find jobs with TNCs soon after receiving their MBA.

Hancock gives in his revealing book a short formula on how the game of foreign aid works:

> This is how the game works: public money levied in taxes from the poor of the rich countries is transferred in the form of foreign aid to the rich in the poor countries; the rich in the poor countries then hand it back for safe-keeping to the rich of the rich countries.[51]

In the view of this author, the way the World Bank development loans work is somewhat different from that mentioned here. The first difference is that nothing is taken from the rich countries. The poor of the poor countries pay for everything. The second difference is that as we are dealing here with loans, although low-interest (7–10 percent) ones, the poor of the poor countries give back more than they receive. Quite often, as the above examples show, poor countries' governments add their own funds to those projects and parts of those are also moved to rich countries. Third, the money is shared among the rich of the rich countries and some rich—those who have access to closed and often nontransparent local project management processes—of the poor countries. Therefore, it is not difficult to see that, under the disguise of development support, resources are moved from poor to rich countries. The example of the World Bank loans for the development of Eastern European higher education systems backs Hancock's argument that, despite the impressive funds spent on foreign aid, the net cash flow is from the South to North. One cannot, however, ignore the fact that for a professor of a developed country's university, acting as a World Bank consultant offers an opportunity to join the TCC; even a local consultant, though considerably less well compensated, takes with such a job a significant step toward the great goal of becoming an international man or woman.

Conclusion

One does not need to be an orthodox Marxist to realize that the means used for development do not achieve the ends. After all, the only positive program the latter group as well as its revisionist comrades have been able to come up with, is a call for a suicide in the consuming fire of the revolution. This, however, does not mean that the critique they offer should be rejected together with the solution. As an OECD report on sustainable consumption indicates, current levels of consumption cannot be spread on a much wider scale,[52] while global culture pressurizes populations into growing consumption. This obviously only widens the gap between consumers and nonconsumers.

In this global struggle, education is caught right in the middle. On the one hand, education has the potential to prepare markets in developing countries for sophisticated and expensive goods and services. That would perhaps justify the attention international financial institutions are paying to developing education in many parts of the world. However, in times when universities in the developed world are turning into aggressively entrepreneurial knowledge industries and are forced to earn ever-growing shares of their budgets through consultancies and other services, patience to develop real markets in developing countries is running short. Profits are to be reaped immediately. Under such conditions, human capital investment programs fail. Available resources are funneled back to the coffers of Western universities and service-providers almost as soon as these are made available to recipients. In the case of loans, interest is added so that development never happens beyond the level of local, often corrupt, elites.

As for the World Bank, one would perhaps be happy if it could succeed even in promoting growth and consumption in the developing world. The reality looks, however, more as if the primary function of its loans is to remove any resources developing countries may have. Robbing the developing world is counterproductive even within its own working paradigm: the global economy needs new markets. It was B.F. Skinner who long ago found how difficult it was to postpone an immediate reward in face of a higher but distant benefit. The result is that for most of the world, the consumer utopia never arrives and the global marketplace is turning into everybody's war against everybody; killing, however, by "other means"—peacefully.

CHAPTER 12
TOWARD A MODEL OF HIGHER EDUCATION REFORM IN CENTRAL AND EAST EUROPE

> Being led by good intentions, we fought on both sides and shot ourselves while defending ourselves from ourselves.
> Anonymous member of the Central Committee of the Hungarian Socialist Workers' Party on the 1956 Revolution, a Paraphrase of István Rév's Paraphrase[1]

The reform of the post state-socialist higher education systems of Central and East Europe has been in progress for more than a decade. Results are, to put it politely, mixed.[2] While the soft touch of the Velvet Revolutions on East European universities has given many a good reason to lament the "stolen revolution,"[3,4] the "peaceful and dignified ways"[5] in which these revolutions were carried out have allowed the higher education establishment to continue without universities losing their status or professors of pseudo-sciences like scientific communism, their dignity.

What is most intriguing about the reforms is that it is increasingly difficult to understand what was so wrong with communist higher education systems in the first place that the need for reform was raised. Gradually, very good communist higher education has given way to excellent *Western-style higher education*; the main difference between the two being only that the Department of History of the Communist Party of the Soviet Union now teaches political science.[6] The logic of the argument seems to lie in the fact that, as the professors are the same, the quality of teaching has not suffered.

Reports from former East Germany reveal that, of the whole of state-socialist East Europe, only the universities of the German Democratic Republic (GDR) were filled with servants of the totalitarian regime and security service agents, and that for the unification of the country and its higher education a major revamp had to be undertaken. While the other countries, including those that were no less controlled by communists and secret police than East Germany, also declare their plans to soon join Western structures, systematic changes in the higher education they offer, remains rudimentary if anything. More often than not, full credit is given to the glorious past of higher learning and satisfaction is expressed with the fact that academia is passing through these difficult times in such a friendly manner.

In a way, this stands in stark contrast with popular demands a decade ago, some of which openly called for the closure of the reproductive structures of the communist *nomenklatura*. It looks as though something dramatic has happened in East

European higher education that has allowed it to reestablish its legitimacy without anything more than cosmetic change. Later I shall discuss the recent reforms in East European higher education in the context of a model developed by Elster, Offe and others for the reform of constitutional legislation in five East European countries,[7] and show how it could come about that for no reason systems should have been preferred to the GDR's higher education provision, getting away with little change while the latter was completely replaced.

The initial calls for radical reform have been replaced by a more "natural" approach. However, the hope that biology will resolve all the problems in East European higher education may well be misplaced, for by the time the old generation of professors retires they may have successfully reproduced themselves. Although natural forces cannot be ignored, particularly in countries where life expectancy has fallen well below 60 years, the deadly forces of alcoholism, nuclear waste, and malnutrition cannot resolve the issues related to teaching and the building of research capacity. The other side of the story is that in some countries living professors do not retire and retired professors do not live. Some of the Russian university leaders declare openly that they are forced to keep the faculty on the payroll, irrespective of age, because forcing a professor to retire with a monthly pension equivalent to $15 constitutes a death sentence.

The Missing Starting Point

Many of the recent reports concerning East European higher education reforms focus on the successful emulation of one or another Western-like structure, apparently under the supervision of omniscient junior civil servants from the European Commission, or the mimicking of processes like strategic planning, the meaning of which remains somewhat distant when daily survival is the ultimate aim. Broader contexts beyond the fact that for some unfortunate reasons—usually external and distant ones—those higher education systems were not Western in the past but are now *catching up*, do not attract too much attention.

Discussions on the Westerness or non-Westerness of East European higher education are largely misplaced and meaningful only as long as "Western" equals "American," which again equals "global."[8] As far as East European higher education has carried an educational mission it has always been Western. What else could it be? *Ex-cathedra* lecturing, the principal sin for which young American friends castigate *communist education*, is unfortunately a very traditional European way of teaching. In his study on Italian higher education, Clark gives a rather good description of life in a continental European university,[9] which—if one excludes party worshipping, direct brainwashing, and collaboration with security services like the Stasi, the KGB, or the Securitate—is not that different from the East European experience. One could almost substitute *the Soviet Union* for *Italy* without doing too much damage to either party. The fact that state-socialist higher education had a strong mission that violated the whole idea of teaching, learning, and free intellectual inquiry, actually turning it upside down, is completely another issue. Furthermore, while significant efforts are being made to introduce active learning methods in East European higher education, the moral corruption of the whole establishment is perceived as irrelevant.

When discussions began in East European higher education, around 1990, about what were perceived as being its main problems insufficient involvement in research, excessive teaching loads, production of graduates for centrally planned economies, and, of course, too low a level of remuneration compared with Western countries were mentioned. By and large, universities distanced themselves from the communist establishment—what it had produced and reproduced for the preceding half a century or more—and blamed external conditions for the compromises it had made and looked for new privileges. This stands in marked contrast with the radical reforms being implemented at the same time in the former GDR. There, faculty members and administrators at all levels had no option but to acknowledge the full consequences of their contribution to the *evil regime*, and vacate their positions. Some of them were hired back, but many were not.

Distancing itself from state-socialism and moving to the camp of free-market liberals or, seemingly more commonly, radical nationalists, was probably the only strategy that would have allowed the university to rapidly generate sufficient legitimacy to allow it to ignore the need for major internal reforms. Only in East Germany was there a sufficiently strong external power to intervene when the labels on the office doors changed overnight from Marxismus–Leninismus to Popperismus, requiring individual assessment of each of the faculty. Fortunately or unfortunately, East Germany had the *Bundesrepublik* that was willing to pay the cost of the change and lend its human resources for the radical replacement in East.[10] In East Europe there was certainly less political pressure, and after initial confusion higher education was able to consolidate its forces against popular challenges.

Even the attack launched by newly established independent and private universities against traditional higher education was only able to present a limited challenge. Initially, hundreds of new higher education institutions mounted a massive attack against the state-run universities, charging them largely with what the West German higher education had charged its East German counterpart, of being a part of the communist political establishment, intellectually dependent and professionally irrelevant. In East Europe, however, traditional higher education and the (new) state quickly rediscovered each other.[11] One major reason for this was the fact that the new political establishment had been produced by the same old higher education establishment. In a sense, any serious questioning of the legitimacy of state-socialist higher education by the new political establishment would have put its own legitimacy at risk. With this strategic partnership—the state legitimizing the university through the newly found *Western* method of higher education management, accreditation, and universities declaring their support of the state and its nationalist goals, independent higher education had been effectively marginalized. Private universities—unrecognized and impoverished—quickly became what in the Brazilian context Levy calls "the demand-absorbing sub-sector."[12]

The second important issue that has helped East European universities to recover from their legitimacy crisis has been the rapidly growing public demand for higher education qualifications. All formerly state-socialist countries experienced significant economic difficulties as the planned economies and production sectors collapsed. Reorientation from production as a mere fulfilment of state plans to the production of useful goods to be sold on the markets meant closing down a large part of the production sector. That has not always been successful. The social cost of those reforms has been high, sometimes unbearably high, and after more than a decade some countries are still trying to balance market reforms against the need to offer job security of some kind. However, recent history also demonstrates that the longer the reforms are postponed the more difficult initiating any change becomes—deterioration of macroeconomic conditions leads to more aggressive rejection of any reform programs that inevitably further threaten the well-being or even survival of many. Balcerowicz explains that discontinuing the production of *pure socialist output*, production that has no value beyond meeting the quantitative targets of the state plans,[13] is valid for higher education as much as it is valid for industry. While one may sympathize with the uncompromising directness with which statements like this address the core problems in East European higher education, its antiintellectualism and failure to see a difference between a shoe factory and a university reaches a level close to that of British higher education consultants.

The first and most visible difference between industry and higher education is that while industry could not sell the products it had produced for the state on the market (stored them in the warehouses for a while and then recycled them), the production of the higher education sector has never met such a market test, and despite the efforts of technocrats to compile lists of skills and knowledge for each field of studies, will hopefully never face such a test. Much of the market value of the degrees universities award is directly related to prestige, which is fairly difficult to express in simple cognitive terms. Advocates of the Oxford collegiality offer the best example of noncognitive benefits of highly elitist education. Consider the following quote from Stephen Leacock's "*My Discovery of England*":

> What an Oxford tutor does is to get a little group of students together and smoke at them. Men who have been systematically smoked at for four years turn into ripe scholars...A well-smoked man speaks and writes English with a grace that can be acquired in no other way.[14]

That makes the failure or success of state-socialist higher education a rather difficult issue to assess. While many who argue for its success present the results in natural sciences and some applied fields like nuclear energy and the space program as their main arguments. Opponents mention the almost shameful poverty in which some of these countries live, unable to manage and use any of their resources. Whether the fact that in 1989 the majority of the graduates who had, during their studies, taken an enormous number of courses related to Marxist–Leninist ideology and then moved to the *capitalist* camp should be counted as an evidence of success or failure of those higher education systems remains a moot point. They had clearly failed to teach what they were expected to teach about the superiority of the communist system and the global communist revolution. The question is whether they taught more or less than that. Obviously, communist higher education failed in its task of molding the new type of personality, the communist man. The current reality shows that instead it produced a selfish and ruthless profit-seeker, a robber capitalist.

There are not too many who could criticize the system from the practical point of view, based on the inability to train high-quality human resources, if one can use the terminology common to modern forced labor, although in some countries at least the usual statements on high educational levels have been challenged by those who complain of a lack of qualified and motivated blue-collar workers. There are even fewer who could challenge the continuation of the bad old ways in higher education on moral grounds. Ten years after the revolutions, a majority of the possible critics are still trained by the system they are expected to criticize. Contrary to Balcerowicz's expectations, the pure socialist output in higher education has not shown any decline. While attempts to modernize higher education have remained largely cosmetic, being limited to the rectors flying to the West and computers being shipped to the East,[15] growth in enrolments of 50 percent or more is not rare among new, as well as not so very independent, states.

The state is not in a position to challenge the traditional higher education establishment because this is the very source of its own legitimacy. The public cannot question it either because for most, higher education constitutes the only possible source of upward social mobility. When the majority of the people were stripped overnight—as a result of shock-therapy of one form or another—of the real or illusory security that the communist states had provided them, higher education remained as one of the few affordable means that could possibly facilitate movement to a better life.

Thus, we have defined the two main characteristics of post-1989 higher education in East Europe: strong public demand for free-of-charge higher education; and large growth in enrolments. Unfortunately, together both set serious constraints on possible reform, often condemning higher education to a long and agonizing decline. However, in times when any higher education qualification is seen as a necessary condition for paid employment, not too much attention is paid to content and quality of education or the ethical standing of the faculty. The result is often similar to what de Moura Castro and Levy describe in the case of Latin America: "The implicit bargain with students is that no work is done and no one complains."[16] Relatively quick and painless restoration of the legitimacy of post state-socialist higher education, amongst others, means that universities have regained a powerful position in society and can dictate the range of possible reforms they will accept. This, however, can still take place within the broader context of post state-socialist social and economic realities. The following three conditions seriously limit how far universities are able to focus on the restoration of their own privileges: (i) extreme shortage of public funds, (ii) massive, often aggressive public demand for publicly funded higher education, and (iii) the inertia of higher education institutions as organizations and the conflicting interests of its internal factions.

When East European higher education systems chose to distance themselves verbally and symbolically from communist ideals during the more or less peaceful revolutions, they were not free to choose the targets into which they would transform themselves. The range of options possible was limited by conceivable and known models. Quite often, the number of possible alternatives varied between zero and one. Developing a new creative higher education model has so far not figured in the policy agenda of any post state-socialist country. One of the reasons for this may be that developing such a model may take more time and resources than has been available. However, universities certainly perceived no need to create alternatives for themselves.

Elster, Offe, et al. recently published a study on institutional reform in post-communist societies that reaches well beyond traditional East European politically predetermined production or Western confusion-cum horror stories. They propose that: "the desirable shape of the new regime, the institutional patterns to be adopted in its consolidation, are deeply contested among the poorly organized proponents of the three divergent orientations. These three orientations or ideas focus, respectively, on the *distant past*, the '*modern*' West, and what is stilled deemed, increasingly, the accomplishments of the *immediate* past of the state-socialist system."[17]

This approach is helpful in understanding the orientation higher education has assumed under the major forces just analyzed. Applied to higher education, I would suggest that while we cannot possibly present the post-1989 period in East European higher education as moving in one single predominant direction, there seems to be a general dynamic in the processes that starts from a rather strong orientation to the distant past moving toward the modern West. The most recent period has, more than anything, been controlled by the immediate past and its romantic reconstruction. In some countries, the immediate past seems to have become the predominant orientation in the late 1990s. Later, I will briefly describe the implications of each of these three orientations for East European higher education. However, East Europe covers a vast region and its countries are increasingly different. Hence, not all the arguments apply to all countries for all times. For example a return to the *distant past* is hardly a viable solution for most of the former Soviet Union countries. It is simply too distant. So, the continuity of the *recent past* seems to be even more strongly present. There are also regions where the Western agendas are at least challenged by other, sometimes radical, programs imported from other sources.

Back to the Roots

In the context of East European higher education a return to the distant past or to historically legitimate models usually means the return to the Humboldtian ideal of the research university. Indeed, the early post-1989 period returned to this model in the thinking of many reformers. A return to the principles of academic autonomy—*Lehr-und Lernfreiheit*—was one of the main aims of the university community of the late 1980s. When the power of the totalitarian state was eroded under the pressure of the Velvet Revolutions, intellectual circles took orchestrated steps toward stripping higher education of utilitarian ends—production of manpower in the quantity and variety defined by the state plans—and a return to Humboldtian innocence. Highbrowed discussions were held on the true meaning of the *Universitas*. Institutes of higher education traditionally responsible for relatively narrow professional training in medicine, engineering, teacher training, or agriculture were generally renamed universities: Polytechnic Institutes became Technical Universities, Pedagogical Institutes were renamed as Pedagogic Universities, and so on. Within the university, departments were divided into Chairs following the idea of the *Lehrstuhl*, expressing the interests and cementing the power of senior professors. Part-time and distance education programs, considered to be low quality and as serving the old ideals of social mobility from peasant youth to young communist leaders, were closed in some places.

Most of all, this expressed the desire of an academia that sought to break free of the straitjacket of training manpower for centrally planned economies. Fewer teaching hours, more security and freedom of research, international mobility, and an income comparable to that of highest-ranking research universities in the West constituted the core of the reforms from the perspective of senior professors. Unfortunately, this was not realistic. In times of economic crisis, governments had neither the means nor the political will to finance the reform of massive higher education systems into elitist groves of academic leisure. While there has been a significant overlap between the two concepts, gradually the idea of introducing contemporary Western models gained ground.

Westernization

A return to models from the distant past serves, at best, the interests of senior faculty. However it neither fits the agenda of the new political elites in relating themselves to contemporary Western structures and access to their resources, nor to public expectations for broader access to higher education. At a time when West European higher education is going through massification and diversification, the idea of restoring the nineteenth-century research university on a massive scale appears particularly controversial and parochial. Not to mention the more recent idea of the entrepreneurial university, which, if moving eastward, would reduce to the bare minimum any expectations of combining mass higher education with academic autonomy and state-guaranteed financial security.

Given the circumstances—lack of means, political conditions, and social support to carry out the plan to transform mass higher education into an elitist system—catching up with Europe/the West/the World becomes the primary cover under which the truly humble goal of East European higher education's survival is to be pursued. According to Elster, Offe, et al.:

> The Western-oriented modernizers base their claim to leadership on a consequentialist reading of the institutional setup of the OECD world. They conclude that representative democracy and market economy yield prosperity, and as prosperity is evidently the highest, at any rate the most urgent priority, all "we" need to do is imitate and transplant western patterns—above all in order to motivate the provision of urgently needed western assistance and cooperation. ...Among the modernizers, the negative assessment of the old regime is this: We want to be a "normal country," and the long period of communist rule has prevented us from becoming one. The old régime must be blamed for its authoritarian inefficiency.[18]

The process includes its own controversies and ironies. For example the Westernization agenda exported by experts from the EU includes elements like establishing a closer relationship between higher education and industry, something the communist regimes had tried, always with rather poor results, throughout their histories. Moreover, dissatisfaction with this particular policy was one of the reasons why East European academics harked back to the distant past models in the first place.

Fortunately, the Westernization of post state-socialist higher education has been significantly supported from the budget of the EU—another good reason to accept it.

The content of Westernization remains, however, obscure. The program has been based on the largely invalid premise of the existence of a number of fundamental characteristics on which the Western and former communist higher education systems oppose each other. Despite many widely accepted stereotypes, both of them remain sufficiently heterogeneous—weakening the relevance of much of the policy advice. Moreover, East European universities cannot be easily penetrated by any external power. Significant gaps continue to exist between official knowledge on how systems and institutions function and the everyday life in government offices and university departments. Therefore, advice that responds to the problems as described in self-promoting reports compiled by local interests groups or in studies produced by external experts, the sensitive details of which are negotiated before publishing (like, for example has been the case with the Council of Europe's advisory missions' reports) fatally misses the target. Significant language barriers, especially in the countries of the former Soviet Union, complicate any direct dialogue and curtail drastically the circle of those having access to the *new* knowledge.

Some of the implications of the Westernization agenda, for example changes in the degree structure or the introduction of quality assurance mechanisms, have been extensively discussed earlier in this book. It would be safe to say that the Westernization program has almost no impact on the students' learning experience. Many international projects that aim at introducing methods of strategic management in East European universities have provided an enjoyable atmosphere for rectors to socialize with each other and the *prominent Western experts*. The fact that ca. 35 million USD spent by the TEMPUS-TACIS in the fSU (former Soviet Union) countries (excluding the Baltic States and including Mongolia) in the years 1993–99 on University Management has not produced a more efficient management model, even on paper, is a remarkable example on impenetrability of the East European structures that Baudrillard so colorfully describes.[19] The other important aspect of this incredible failure is the inefficiency, if not corruption, of Western aid itself.

Continuity and the Japanese Solution

The *recent past* is the third reference point Elster, Offe, et al. argue, to which post state-socialist countries in Eastern Europe relate their institutional reforms. In higher education, it is so not so much expressed through reform as the lack of it.

Current East European higher education reform discourse seems to ignore systematically the impact of society on higher education. It is somehow expected that whatever the political and economic situation may be, higher education is detached from it, acting from the position of integrity, wisdom, and altruism. Increasingly the same view is retrospectively applied to state-socialist higher education. While much of the current *transitology* literature, particularly that published by Western scholars, demonstrates the transition from *nomenklatura* to *kleptotura*—referring to the continuity between the corrupt ruling elite then and now—still a naïve view seems to be gaining ground that higher education has not been part of those unfortunate developments.

When it became clear that academic privileges from before the Second World War or earlier could not be restored, East European higher education has drifted. Under growing external pressures, it has sought to protect itself against further changes. Agencies like the World Bank often tie their loan programs to job cuts and mergers of institutions. As those reforms put more at risk than simply the faculty's modest living standard, it comes as no surprise that universities are not particularly keen to cooperate in, let alone initiate, reform. The university can only fight back, becoming more conservative still. One of the arguments that helps to erect a protective wall against change, involves summarizing on the past success of communist higher education, the days when the leaders of Third World countries were trained in Moscow or Budapest, how Soviet scientists built Sputnik and nuclear weapons (actually stolen from the British), and constructed a supersonic passenger aircraft (a rather unsuccessful copy of the Concorde), and so on and so forth. The aim is to restore the legitimacy of higher education and to avoid changes that threaten the institutions and the people within them. Unfortunately, the lack of active higher education policy is already showing consequences that may well be worse than those of a well-thought through policy, although the opposite could also be true: a nonintervention policy may be the most efficient way of solving difficult problems at a time when consensus behind any possible policies is missing, as was the case for Japanese agriculture after the Second World War.

There are situations and times when pursuing any active policy most probably stops serious social unrest. In an article, Drucker described the situation in Japanese agriculture after the Second World War when some 60 percent of the population tried to make their living out of it and were "utterly unproductive." Having a choice between either risking serious social disruption as a consequence of a policy "to move off the land or become more productive" or doing nothing, the Japanese bureaucracy chose the latter. While disastrous from an agricultural point of view, the policy of no policy—quite similar to that of post-1989 East European higher education policy—was a huge social success: "Japan has proportionately absorbed more former farmers into the urban population than any other developed country without the slightest social disruption."[20]

The situation of many of the East European higher education systems is alike to postwar Japanese agriculture. Higher education serves as a major social safety valve. Providing free higher education keeps large numbers of young people off the streets when job opportunities are limited. Avoiding any job cuts in higher education and operating the systems with faculty/student ratios as low as 1:5, with remuneration close to or slightly below the subsistence minimum prevents the worst from happening—angry unemployed professors demonstrating on the streets. Obviously, the first side effect is that, as with Japanese agriculture, those who can move out of the sector do so, particularly the younger generation. The no-policy policy also defines the range of possible changes the system can absorb. It actually transforms the largest part of public higher education, into the last resort of the least capable professors and economically underprivileged students. The best faculty either move abroad, leave higher education, or find better remuneration at private higher education institutions. Students have largely the same options.

At the same time, the faculty is irreversibly ageing. The dilemma universities face concerning personnel policy is evident: recruiting young faculty reduces efficiency even further, at least as long as the older generation has to be kept on the payroll for humanitarian reasons. It is also not rare to find the older generation taking a more active position by blocking the younger generation from entering academia with arguments about high academic standards and bureaucratizing the process of awarding higher degrees, just as it was once done in the Soviet Union.[21] The lack of a more active policy of faculty development is sometimes justified by the fact that a large number of current faculty will leave the universities during the coming decade anyway, resulting in more balanced faculty/student ratios and opening some job perspectives for the younger generation. This position fails to recognize, however, that two decades of drifting will seriously damage the academic mission of the universities as they are transformed into old people's homes. When this mission is finally completed it will not be long before a new round will need to begin, the old people's homes will have to be transformed back into universities under equally difficult economic conditions.

The no-policy policy is, in its ultimately passive form, a difficult one to carry out consistently. As with any other policy, it needs to be justified. The ground on which it can be defended in Eastern Europe is the perceived or constructed success of its higher education in the past. It is obvious that the university community is vitally interested in presenting its great achievements in their full or even greater glory. However, the nations themselves need to present their intellectual success stories rather than the instances of ideological manipulation and rule by foreign governments. The newly independent nation-states in the Baltic, Central Asia, and the Caucasus need to construct their intellectual histories for the sake of nation-building, and for this they turn to the universities. The Central European nations also do the same to overcome what is perceived as a period of national shame under the rule of "barbarians from the East."

The social role higher education plays in the current conditions, as the last remnant of the communist utopia, may well be more important than changing it at a time when the future of the economies and the structure of the new labor markets is unclear. Not doing anything, however, does not mean an absence of change in higher education or, as Offe explains, "In this sense, an abstention from a politics of regulation is 'politics' nonetheless—as much as in intentions as in its consequences."[22]

Yet, this stance erodes mass higher education to the extent that public universities are losing their academic functions, so much so that one of the few things relating them to higher learning is the fact that they still award degrees. Quality education providing access to good jobs is concentrated in a limited number of leading *national* universities, and private elite institutions. The new system, with its low-quality demand-absorbing mass sector and high-quality elite institutions, is one of remarkable social injustice and inequality—something that stands in stark contrast with the longed-for utopia that is its main cause. The final outcome of the no-policy policy agenda in East European public higher education may well be an echo of what happened to Japanese agriculture after the Second World War—it will disappear. While this was precisely the desired outcome in Japan, it may not be so in Eastern Europe.

The return to the recent past is the next-best solution for a large part of East European academia after failing to restore the Humboldtian ideal. The remnants of the Westernization discourse should be considered as a part of an international fund-raising exercise. Baudrillard expressed his own view on this some years ago: "Now, contrary to the apparent facts which suggest that all cultures are penetrable by the West—that is, corruptible by the universal—it is the West which is eminently penetrable. The other cultures (including those of Eastern Europe), even when they give the impression of selling themselves, of prostituting themselves to material goods or Western ideologies, in fact remain impenetrable behind the mask of prostitution. They can be wiped out physically or morally, but not penetrated."[23]

Searching for the Source of Change

After wandering for a decade in the wilderness of post-communist transition, East European higher education has returned to its starting point. Not unlike the Jews in the desert who told Moses that Egypt was not such a bad place after all. From a distance, the meat-pots and garlic it offered to obedient slaves looked more attractive than manna everyday. Even the graves of the less cooperative slaves were better than they would have with their bones spread in the desert sand. East European professors remember the good old days when the state paid the bills and asked very little in return.

Only 14 years after the closure of the last forced labor camps in GULAG, it almost seems as if university professors had never sent their colleagues there by writing reports on those who had, in their naïveté, misunderstood the communists' statements on free speech and used their typewriters to copy silly political statements. Surely, such unreasonable behavior should not be tolerated in the house of reason. Writing a report on a troublemaking student posing a *provocative* question, for example on the status of agricultural productivity in the Soviet Union, was also considered normal. After all, one should understand where to speak and what about. According to Soviet psychiatric practices, an inability to understand the obvious benefits of the communist system was considered a symptom of mental disease.[24] Even if common sense should have dictated that the communist system might have had certain problems, and not only for those suffering of schizophrenia, not understanding one should not mention such things in public was certainly sick. So the reports were written. Most of the people who wrote them remain in their universities to this day, almost everywhere except in the former GDR.[25]

For many of the faculty, the situation as it was under the old totalitarian political regimes was the next-best option after the Ivory Tower in which they could have enjoyed their alchemist experiments without too much scrutiny or accountability. The Ivory Tower project did not work out for three main reasons: first, none of the East European postrevolutionary governments, with Slovenia as the only possible exception, had the resources (having often lost up to a half of their industrial output) to meet the universities' expectations for funding. Second, the public, having lost the illusion of security the communist regimes had offered, did not share the idea of elite higher education. They demanded massive access, not so much for the development of knowledge and skill as for credentials that would facilitate upward social mobility, particularly for those who did not have social capital that could be exchanged for state assets through privatization. Third, many faculty had found, much to their surprise, that the foreign agencies did not really appreciate their reform program and that the Humboldtian ideal did not sell, neither to the World Bank nor to the European Commission. What these agencies wanted to accomplish felt surprisingly close to what the communists had tried for decades with mixed results—to promote cooperation between university and industry, to vocationalize or at least professionalize higher education, and to measure its economic efficiency. Having understood this, many of the faculty actually began to sabotage the reform initiatives supported by foreign organizations. The never-ending, or better still, never-beginning story of the World Bank project in Hungary, launched several times since the early 1990s, offers a paradigmatic example of how academia can effectively block reforms, maneuvring skillfully on the political scene, manipulating public opinion and political interests.

Foreign aid has only an indirect connection to what common sense may consider its primary mission. It has its own life and logic, which may be related more to the interests of provider organizations and service-providers than to the needs of the recipients. At least in East European higher education, the lion's share of aid funds goes to Western experts for their consultancy and expertise. The remainder corrupts local elites, takes rectors and their international relations officers—generally acting as interpreters—to London or Loewen for an *exchange of experience*, and brings a few PCs to the campuses.

Misunderstanding between donors and recipients over the past decade has been monumental. Western experts have spent much of their time and available resource in trying to understand the reality of those they are supposed to help. They have not got very far. For example the persistence with which Western experts repeat, one after another, the misconception that higher education institutions in the Soviet Union were not involved in research is amazing, because everybody who has ever worked in that system knows from personal experience that one-third of the faculty's work time was devoted to research. Furthermore, there were research associates who were involved only in research. After a decade, the first meaningful results of a diagnoses on East European higher education systems are emerging. A recent OECD report on Russian higher education is, in its careful way, trying to address some of the major problems the system faces.[26]

Some of the sincere attempts by the West to help the East have been remarkably naïve, seemingly based on two flawed premises: first, that East European faculty are ignorant and; second, that they are aware of their ignorance and earnestly wish to overcome that deficiency. There is an obvious conflict here. Professors in Eastern Europe, as much as anywhere, think that they are genuinely smart. When a well-meaning agency sends doctoral students to enlighten professors, the latter simply ignore the whole exercise. One could imagine that even in less status-sensitive societies than those of Eastern Europe, professors would react similarly.

There is one other reason, in addition to the apparent ignorance of Western experts on matters of East European higher education and their underestimation of the cognitive abilities of the faculty there. There is a certain amount of skepticism within the East European higher education community about the seriousness of aid attempts and a perceived cultural superiority, particularly in comparison with the Americans. This is by no means unique in Europe. It is often perceived that Americans, even those coming with the best of intentions, fail to see their own limitations. An example is when Gellner writes of the American Declaration of Independence, "...the things which they hold to be self-evident were unintelligible, blasphemous, heretical, or proscribed for the large majority of mankind." On the American perspective of history he comments, "They have some sort of vague notion about George III, but the whole of European history between the Pharaos and George III is one big mess, which is interesting for tourism but otherwise does not terribly interest them. America was also born individualist, liberal, and rational. ... But it means that the problem is invisible to them because the decencies are taken for granted."[27] When accepting Western money, East European higher education is convinced that not only is it the duty of the West to produce it, but also that the West is not really aware of their rich traditions or the level of academic excellence achieved under the communist regimes. This, many people seem to believe, gives them the moral right to accept the money and use it however they wish, which may or may not correspond with the *approved project description and budget*.

East Europeans are well aware of the dual morality that communist education created. Even so, they still think that by-and-large they are the second smartest people on the planet after the Japanese. This is only one rung lower than the one insisted upon by the formerly ruling comrades. There is every good reason to believe that the communist ideology according to which birth in a communist country makes one a superior being, is still alive and well. It may well be that this can be passed by birth to the next generation, even in a capitalist environment. This could be supported by Mitchurin's and Lysenko's teaching on the inheritance of acquired characteristics. For our case, it means that two or three generations under communist rule have had an irreversibly positive effect on our genomes, something that Westerners lack. While the reasoning behind this is manifold, there is a prevailing view among East Europeans that there is not much the West could contribute intellectually, though one should not reject the material goods it can offer.

The logic of post-communist superiority may well be that *although we shot some of ourselves who happened to be slightly less ourselves than we are, our intentions were inherently good, and finally, we did a lot of good things, and most importantly, we produced and read more books than anyone in the West*. This is ignored by well-intentioned Western agencies who send graduate students to teach professors in Eastern Europe. No wonder the programs lack credibility!

Ten years after the revolutions, it seems evident that higher education has not, on the whole, been reformed. Some governments, Romania for example are talking about starting genuine reform now. Others, like Russia, seem to be convinced that their higher education is still the best in the world. What it really needs is just a bit more funding. The outstanding question is whether, given the continuity in East European higher education and its skepticism about the West, there is another possible source of reform in addition to that of local academia (which has no interest in changing), governments (which are caught in a political limbo unable to produce more than rhetoric on reform policies), and the so-called international community with its limited credibility and entrepreneurial approach (in the worst possible sense of the term) to the whole affair.

Many share the conviction that there is no source for reform. The labor market is filled with under- or unemployed graduates. Long-term labor-market demand is unclear. For universities no change is the second-best position next to the impossible, and students are not as interested in skills and knowledge as they are in credentials. It is East European cynicism to say that what universities do is irrelevant anyway, a degree is the only thing that matters. In a way, Lajos was right in suggesting, "H(igher) E(ducation) is unable to reform itself" and that "governments are too weak to manage a deep reform at all levels of HE because such management could endanger the values and functioning of the system."[28]

His proposed solution is strikingly simple. Access to higher education should be opened to all applicants. The masses of revolutionary youth then entering the universities would force the corrupt establishment to change. One could raise two possible arguments against such a proposal: the first is ideological. Lajos's proposal has a long history. Trotsky came up with a similar idea back in 1918, but failed. In the hope of releasing the creative initiative of the *revolutionary masses* and setting off the World Revolution, he offered the following formula to the newly emerged Soviet Russia for peace negotiations in 1918, "neither war nor peace but the Army to be dissolved." The other argument is practical. Open access does not apparently reduce corruption or force universities to change, but corrupts them still further. Bringing a large number of unprepared students to the university following public pressure and the need to generate funds through charging the students admitted above the Ministry's admission quota a fee, will not drive the system toward higher quality and further relevance. The bottom line of the latter is that higher education is relevant as long as a more or less valid document certifies it. A leading *national university* in the capital of a Central Asia country reportedly admitted 1,500 students for business studies in 1999, whilst having no intellectual or material resources to educate them. However, those involved, faculty and students, are well aware that paying the fee alone will bring the degree in five years. There are many who say that paying a somewhat higher fee would lead to the same degree without wasting all those years. Unfortunately, a large majority of the revolutionary youth is no less corrupt than the establishment itself. Students are no less willing to offer cash for a good mark than the faculty are to accept it. Parents run themselves into the ground in search of the friends-of-friends who can be bribed to secure admission. Finally, everybody is well aware that the professors of history of the Communist or Socialist Party cannot teach political science, but they still do.

Such moral corruption and the persistent dual morality threatens East European higher education most, for the simple reason that anybody raising the issue is shooting themselves in the foot. The moral commitment of intellectuals is silenced in Eastern Europe. Its historical success has been retrospectively constructed on perceived success in certain purely technical areas. Rorty, however, argued against any specifically epistemic strategy in science, saying that there is no such thing as a scientific method and that the whole business of science is about moral virtue.[29] As much as this moral virtue was lacking in communist higher education's declaration of the eternal truth discovered by Marx, Lenin, Stalin, Mao, Brezhnev, and so on, we still have a problem even if sociology professors know perfectly well how to calculate correlation. Moreover, once the issue is ignored in its totality, there is no source from which it could possibly be restored. At best, Western experts talk about "managerial effectiveness." One may doubt whether the universities there are significantly more moral than elsewhere. As the wise men from the West have found, truth may actually be economically somewhat counterproductive, if not constituting a source of danger, even if to nothing more than economic well-being. Newton-Smith compared science with other areas of human activity, finding that: "One is not supposed to tear out those pages of one's laboratory notebook that go against the hypothesis one has advanced in print. Clearly, this norm serves the epistemic ends in science. And it highlights a contrast with other institutions such as politics or diplomacy. In the case of these institutions the suppression of data is often seen as a positive virtue."[30] This statement perfectly spans East and West, past and present. Thereafter, any mention of the dual morality of communist societies would be utterly redundant. One would be in a rather difficult position in arguing why a professor of Marxism–Leninism— a warrior in the World Proletariat Revolution and at the same time a politician, diplomat, and ideologue—should have applied higher moral standards than those promoted by a *Western* philosopher. In the days of global capitalism, Chomsky's uncompromising declaration that, "it is the responsibility of intellectuals to speak the truth and expose the lies"[31] remains but a voice in the wilderness. The argument that we do not have access to what could be called the ultimate truth cannot justify uttering things we know to be lies. Applying relativism as an absolute could have catastrophic global consequences for the academic world, which would then require only a politically correct justification to lie—the communist cause, the capitalist cause, the monarchist cause—anything, even the open society cause, would go, the only thing missing being the intellectuals.

Bourdieu is doubtless right to argue that *homo academicus* always has a double identity:

> Academics (and, more generally, the members of the dominant class) have always been able to afford to be at once infinitely more satisfied (specially with themselves) than we would be expect from an analysis of their position in their specific field and in the field of power, and infinitely less dissatisfied (with the social world) than we would expect from their relatively privileged position. Perhaps this is because they retain certain nostalgia for an accumulation of all the forms of domination and all the forms of excellence....[32]

There have been times and places where academics have been able to accumulate power whilst fulfilling their intellectual duty, but state-socialist Eastern Europe certainly was not an example. In Eastern Europe, being an intellectual equalled being a dissident. Nowadays, many present themselves as dissidents retrospectively. Were one to apply Žižek's definition of dissidence as a public act,[33] many would not qualify. For state-socialist Eastern European academics, and indeed academics in many places, the state acted as the font of symbolic power,[34] which radically excludes their contribution as intellectuals that would have stood in direct opposition to the goals of the state.

Perhaps Western universities have seen better days when academics were rewarded for their contribution as intellectuals. Such times are quickly coming to an end. Kwiek offers a sobering description of the current situation of the university: "The university, surely enough, still functions, but in an increasingly different manner: it either refers to the logic of production and consumption (of knowledge), that is to say, it sells its production with better or worse results, or it struggles violently with the state that is generally, all over the world, less and less willing to support the public university that in turn refers to its rights gained in modern culture."[35] Academics continue to fight for their privileges, arguing that their social role as public intellectuals is important, not having noticed, or not wanting to notice that times are changing. As universities become a part of the productive force, academics have been downgraded from their former position as intellectuals to the level of knowledge workers, that is wage laborers in the capitalist economy. This is perhaps what Kwiek means by the "death of the intellectual."[36] Their work is now measured in terms of the saleable products produced: papers and books written, patents obtained, and so on and so forth. These products may or may not have an intellectual value. The ultimate requirement is that they sell. Therefore, it comes as no surprise that East European higher education reforms over more than a dozen years have had only a marginal impact on restoring the intellectual mission of the university. Western higher education with its ever-strengthening cash nexus is not in a position to lead Eastern Europe in this direction. East European academics themselves, however, are not in a position to address their real weakness—the lack of intellectual commitment. The main difference between Eastern and Western academics is that while East European academics died as intellectuals with the establishment of communist regimes, Western academics are suffering nowadays, under the pressure of global economic competition—everybody's war against everybody.

Conclusion

The return to the recent past, after the failure to return to the Ivory Tower and the rejection of Westernization, has sent East European higher education into free fall. Societal demands for massive distribution of free higher education qualifications, preferably to each according to their need; poverty and corruption, fusing with a romantic reconstruction of "the good old days" has created a completely new equilibrium for years to come. Much like anywhere else, we see a growing diversity of higher education providers offering everything from highly elitist and high-quality education to basement offices selling unrecognized degree certificates. What is missing is control and guidance by professional communities and credible academics over what is going on in higher education. Students, as well as employers are confused about whom to trust and whom not. Worst of all, the mainstream—the traditional public universities and *institutes* are becoming state-accredited diploma mills, spreading a particular message about the post state-socialist state as well as its reproductive agents. East European higher education faces massive moral problems and, for various reasons, Western counterparts are not in a good position to help. One could even argue that they lack the moral authority to do so. One can but wonder how such higher education systems can carry an intellectual mission of any kind.

Epilogue: The Unholy Trinity of Prince, Prophet, and Philosopher

> Sailing down behind the sun,
> Waiting for my prince to come.
> Praying for the healing rain
> To restore my soul again.
>
> Eric Clapton[1]

In the preface to "The Conflict of the Faculties" Immanuel Kant writes, "An enlightened government, which is releasing the human spirit from its chains and deserves all the more willing obedience because of the freedom of thought it allows, permits this work to be published now."[2] The exchange of letters between Kant and the office of His Majesty King Frederick William that followed shows, however, a certain level of tension between the prince and the philosopher over certain of the latter's views on some fairly sensitive issues, such as religion. At one point in 1794, the king's officer ordered in the monarch's name, "We demand that you give at once a most conscientious account of yourself, and expect that in the future, to avoid our highest disfavour, you will be guilty of no such fault, but rather, in keeping with your duty, apply your authority and your talents to the progressive realisation of our paternal purpose."[3]

A philosopher's life has never been an easy one, independent of whether in the service of a prince or some other authority. When serving a prince, the philosopher has to praise him for his great wisdom or face the danger of falling from favor. Executing heretical philosophers has always been a necessary element in the consolidation of power of any despot. Perhaps, the market is a more humane authority than earlier autocratic rulers: it does not execute those who do not agree with its programs. Instead, it permits philosophers who refuse to deliver growing amounts of increasingly excellent knowledge, to starve to death.

This is not to romanticize the days of princes. It is ironic that for the powerful the power they have is never enough. They also require the recognition of their wisdom whether or not they have it. This is not only a matter of history-book royals. Stalin desired recognition just as much—as the first linguist of the Soviet Union and Brezhnev, who years earlier had been a laughingstock for old Bolsheviks like Khruschev, wished to be a literary genius. Needless to say, both of them found philosophers in their courts who knew how to meet the deepest desires of their princes. Under the market regime things are even easier, if the printing costs can be covered almost anything can be published, after which it is only a matter of more cash to find entrepreneurial academics to legitimate the production.

The relationship between money, power, and wisdom is complex. More often than not, perhaps to the shame of philosophers, all three seem to be mutually interchangeable. In a recent address Neave asked, as if we had a choice, "How far is the freedom of learning and inquiry compatible with the Prince—or the market?"[4] Usually, there is no choice. The difference is that, while a prince may or may not be enlightened, the market is always blind. Rich nations are able to establish enclaves where useless, often provocative thoughts can be thought, suffice alone to think of someone like Noam Chomsky at the MIT. The poor cannot afford that, for their daily bread they deliver whatever the market demands: enlightenment certificates, MBA degrees, or field reports for Western consultants.

A dozen years ago, the nations of Eastern Europe drove their ignorant princes away. In many countries, philosophers played a significant role in mobilizing the masses against despots who were not able to appreciate the depth of their knowledge. Soon it became apparent that the market cared even less about it and that waiting for an enlightened prince—or an investment banker—takes much longer than expected. There have been princes, here and there, some even remain, but never again have they been enlightened. Even the wealthiest of men and organizations have not had sufficient resources to cover even a small part of the needs of half a million East European professors, not to mention their students'. Time is running out and such days are reminiscent of those of Jeremiah, "The harvest is past, the summer has ended, and we are not saved."[5]

Attempts to attract the savior from his hiding place have become desperate. Politicians and philosophers produce treatises on the great achievements of the universities and as time passes, the gap between the "*pays politique*" and the "*pays real*"[6] only grows. The more the sector deteriorates the more impressive are the achievements that find their way to the "official" records. The belief is that by repeating the prophecy sooner or later it will fulfil itself. It is a misplaced hope. It is hardly possible that in current circumstances even a very enthusiastic political discourse has the strength to coax reality out from its resting place in "exponential stability" for the simple reason that the prophet lost his credibility long ago, by compiling reports to the Central Committee on nonexistent achievements.

We do not know when the bubble will burst or what the consequences might be. It may well be the case that it will never burst. The former highly centralized East European higher education systems may be cooling out infinitely. These countries do not possess the resources to guarantee the quality of education for more than 3–5 percent of the total student body, those who study at the premier national universities. The rest covers an extremely wide spectrum of activities of high quality, low quality, and no quality whatsoever. But even this is not such a big problem. Ultimately, the market gets what it asks for. Sooner or later there will also be information available on what is what. League tables have already been developed for Russian universities.

The real issue concerns philosophers, the relatively small number of intellectuals without whom society cannot function. Under the previous regime these people, as a rule, were not to be found in universities. Quite often, they were among the blue-collar workers, exercising their freedom at nights as they read banned literature that circulated through underground channels. Under the new regime, they are still not in the universities. The universities view them as useless. They create far too much trouble and generate no cash. Whether East European societies will eventually be able to propagate the necessary number of critical intellectuals and, if they do, whether the university, against all odds, will be willing to accept them, remains a point still to be settled.

Notes

Chapter 1 From Lenin to Digital Rapture: The Everlasting Transition in East European Higher Education and Beyond

1. Paper presented at the 23rd Annual Forum of EAIR—The European Higher Education Society, University of Porto, Porto, September 9–12, 2001.
2. C. Castoriadis, *The Imaginary Institution of Society*, Cambridge, Massachusetts: The MIT Press, 1987, pp. 121–122.
3. J.D. Barrow, and F.J. Tipler, *The Anthropic Cosmological Principle*, Oxford and New York: Oxford University Press, 1986, p. 21.
4. Ibid., p. 23.
5. Castoriadis, *The Imaginary Institution*, p. 111.
6. S. Žižek, *Did Somebody Say Totalitarianism?: Five Interventions in the (Mis)use of a Notion*, London, New York: Verso, 2001, p. 154.
7. K. Marx and F. Engels, "Manifesto of the Communist Party," in Karl Marx and Frederick Engels, *Economic and Philosophic Manuscripts of 1844 and the Communist Manifesto*, New York: Prometheus Books, p. 212
8. E. Todd, *The Final Fall: An Essay on the Decomposition of the Soviet Sphere*, New York: Karz Publishers, 1979.
9. From Nina Rogalina's contribution to the roundtable discussion S.S. Sekirinskii, "Sovietskoe Proshloe: poiski ponimaniia" (Soviet Past: Search for Understanding), *Otechestvennaia Istoriia* (Fatherland's History) 4, 2000.
10. R.K. Merton, *Social Theory and Social Structure*, New York: Free Press, 1968.
11. See e.g. G. Soros, *Open Society: Reforming Global Capitalism*, London: Little, Brown and Company, 2000.
12. Marx and Engels, "Manifesto," p. 231.
13. Ibid., p. 230.
14. D. Lane, *The Rise and Fall of State Socialism*, Oxford: Polity Press, 1996, p. 194.
15. C. Gaddy and B. Ickes, "Russia's Virtual Economy," *Foreign Affairs*, September/October 1998.
16. S. Bruckbauer, "Ranking Productivity and Its Dynamics: Faltering Growth of Eastern Europe," *Economic Trends*, 3 (2000), pp. 29–33.
17. World Bank, *Constructing Knowledge Societies: New Challenges for Tertiary Education. A World Bank Strategy*, Washington, DC: The World Bank, 2002.
18. J. Salmi, "Globalisation and the Knowledge-Based Economy Pose Challenges to All Purveyors of Tertiary Education," Tertiary Education—News Item, *Times Higher Education Supplement*, December 3, 1999.
19. E. Davis, *Techgnosis: Myth, Magic and Mysticism in the Age of Information*, London: Serpent's Tail, 1999, p. 3.
20. Ibid., pp. 259–260.
21. A. Jackman, in *Techgnosis*, p. 260.
22. M. Poster, "Cyberdemocracy: The Internet and the Public Sphere," in David Trend (Ed.), *Reading Digital Culture*, Malden, MA: Blackwell Publishers, 2001, p. 260.
23. A. O'Hear, *After Progress: Finding the Old Way Forward*, London: Bloomsbury, 1999.
24. Jackson, *Techgnosis*, p. 261.

25. R. Coyne, *Technoromanticism: Digital Narrative, Holism, and the Romance of the Real*, Cambridge, Massachusetts and London, England: The MIT Press, 1999, p. 4.
26. H. Murakami, *The Wind-up Bird Chronicle*, London: The Harvill Press, 1998, p. 467.
27. M. More, *The Extropian Principles: A Transhumanist Declaration*, version 3.0, available on the Internet: http://www.extropy.com/extprn3.htm.
28. Murakami, *The Extropian Principles*.
29. I. Havel, *Living in Conceivable Worlds*, paper presented at the First World Congress of Paraconsistency, Ghent, July 30–August 2, 1997.
30. Jackson, *Techgnosis*, p. 121.
31. S. Lem, *The Star Diaries*, San Diego, New York, London: A Harvest Book, 1985.
32. J.C. Lilly, *The Scientist: A Metaphysical Autobiography*, Berkeley, CA: Ronin Publishers, 1997, p. 150.
33. W.H. Newton-Smith, "The Origin of the Universe," in P.J.N. Baert (Ed.), *Time in Contemporary Intellectual Thought*, Amsterdam, Lausanne, New York, Oxford, Shannon, Singapore, Tokyo: Elsevier Science B.V., 2000.
34. Ibid., p. 57.
35. Ibid., p. 65.
36. Coyne, *Technoromanticism*, p. 279.
37. Ibid., p. 111.
38. D. Shenk, *Data Smog: Surviving the Information Glut*, New York: HarperEdge, 1997, p. 211.
39. Ibid., p. 199.
40. A. Vostrikov, contribution to the Salzburg Seminar Universities Project's Russian Symposium, Salzburg, Austria, April 15–19, 2000.
41. F.A. Hayek, *The Intellectuals and Socialism*, London: ILEA Health and Welfare Unit, 1949/98, p. 18.
42. M. David-Fox, *Revolution of the Mind: Higher Learning Among the Bolsheviks, 1918–1929*, Ithaca and London: Cornell University Press, 1997.

Chapter 2   Higher Education Reform in Romania: Knocking on Heaven's Door

1. I. Iliescu, opening speech given at the international symposium "Central Europe—South-Eastern Europe: Inter-regional Relational Relations in the Field of Education, Science, Culture, and Communication," Bucharest, Romania, April 19–22, 2001.
2. J. Sadlak, "The Emergence of a Diversified System: The State/Private Predicament in Transforming Higher Education in Romania," *European Journal of Education*, 29, 1 (1994), pp. 13–23.
3. S.D. Roper, *Romania: The Unfinished Revolution*, Amsterdam: Harwood Academic Publishers, 2000.
4. Romania, *The National Medium-Term Development Strategy of the Romanian Economy*, Bucharest: Government of Romania, 2000.
5. Ibid.
6. D. Sapatoru, *Public or Private? Post-Secondary Education Choices in Romania*, manuscript.
7. Romania, *The National Medium-Term Development Strategy*.
8. Romania, *Declaration*, Bucharest: Government of Romania, 2000.
9. Romania, *Government Program*, Bucharest: Government of Romania, 2000.
10. Ibid.
11. Ibid.
12. S. Bruckbauer, "Ranking Productivity and Its Dynamics: Faltering Growth of Eastern Europe," *Economic Trends*, 3 (2000), pp. 29–33.
13. Ibid.

14. Sadlak, "The Emergence of a Diversified System."
15. Sapatoru, *Public or Private*.
16. Romania, *Carte Blanche of the Reform of Education in Romania*, Bucharest: Ministry of Education, 1995.
17. Sadlak, "The Emergence of a Diversified System."
18. *Carte Blanche*.
19. OECD, *OECD Reviews of National Policies for Education: Romania*, Paris: OECD Publications, 2000.
20. Sapatoru, *Public or Private*.
21. Ibid.
22. Sadlak, "The Emergence of a Diversified System."
23. Sapatoru, *Public or Private*.
24. Sadlak, "The Emergence of a Diversified System."
25. *Carte Blanche*.
26. Romania, *The Law on the Accreditation of Higher Education Institutions and the Recognition of Diplomas*, Bucharest: The Parliament of Romania, 1993.
27. M. Korka, *Strategy and Action in the Reform of Education in Romania*, Bucharest: Paideia, 2000.
28. Romania, *Higher Education in a Learning Society: Argument for a New National Policy on the Sustainable Development of Higher Education*, Bucharest: The Ministry of National Education, 1998.
29. Korka, *Strategy and Action*.
30. Sapatoru, *Public or Private*.
31. *Higher Education in a Learning Society*.
32. L. Nicolescu, *Private vs. State Higher Education in Romania: The Business Community Perspective*, unpublished manuscript.
33. Ibid.
34. Ibid.
35. Ibid.
36. Korka, *Strategy and Action*.
37. D.C. Levy, *Higher Education and the State in Latin America: Private Challenges to Public Dominance*, Chicago: The University of Chicago Press, 1986.
38. *Higher Education in a Learning Society*.
39. Korka, *Strategy and Action*.
40. B.R. Clark, *Places of Inquiry: Research and Advanced Education in Modern Universities*, Berkeley and Los Angeles, CA: University of California Press, 1995.
41. C. de Moura Castro and D.C. Levy, *Myth, Reality and Reform: Higher Education Policy in Latin America*, Washington, DC: Inter-American Development Bank, 2000.
42. *Higher Education in a Learning Society*.
43. Korka, *Strategy and Action*.
44. L. Boia, *History and Myth in the Romanian Consciousness*, Budapest: Central European University Press, 2001.
45. World Bank, *Anticorruption in Transition: A Contribution to the Policy Debate*, Washington, DC: The World Bank, 2000.

Chapter 3 Thirteen Years of Higher Education Reforms in Estonia: Perfect Chaos

1. A significant part of this chapter has been previously published under the title "Higher Education Reform in Estonia: A Legal Perspective," *Higher Education Policy*, 14, 3 (2001), pp. 201–212. Reprinted with the kind permission of Palgrave Macmillan Journals.

2. M.S. Gorbachev, *Perestroika i novoe myshlenie dlia nashei strany i dlia vsego mira* (*Perestroika and New Thinking for Our Country and for the Whole World*), Moscow: Politizdat, 1987.
3. E. Todd, *The Final Fall: An Essay on the Decomposition of the Soviet Sphere*, New York: Karz Publishers, 1976.
4. J. Elster, C. Offe, and U.K. Preuss, *Institutional Design in Post-Communist Societies: Rebuilding the Ship at Sea*, Cambridge: Cambridge University Press, 1998.
5. A. Watson, *Legal Transplants: An Approach to Comparative Law*, second edition, Athens and London: The University of Georgia Press, 1993.
6. *United States-Soviet Relations: 1988, Hearings Before the Subcommittee on Europe and the Middle East of the Committee on Foreign Affairs, House of Representatives, One Hundredth Congress*, vol. I, Washington: U.S. Government Printing Office, 1988.
7. V. Tomusk, "Developments in Russian Higher Education: Legislative and Policy Reform within the Central and East European Context," *Minerva*, 36, 2 (1998), pp. 125–146.
8. For a more precise description of Estonian higher education see V. Tomusk, "Estonia: Higher Education System," in G. Neave et al. (Eds.), *Complete Encyclopaedia of Education/Encyclopaedia of Higher Education on CD-ROM*, London: Elsevier Science, 1998.
9. V. Tomusk, "'Nobody can Better Destroy Your Higher Education than Yourself': Critical Remarks About Quality Assessment and Funding in Estonian Higher Education," *Assessment & Evaluation in Higher Education*, 20, 1 (1995), pp. 115–124.
10. V. Tomusk and A. Tomusk, "Teaching Psychology, Estonia: USSR Revisited," *Teaching of Psychology*, 20, 3 (1993), pp. 175–177.
11. V. Tomusk, "Recent Trends in Estonian Higher Education: Emergence of the Binary Division from the Point of View of Staff Development," *Minerva*, XXIV, 3 (1996), pp. 279–289.
12. ESSR, *Lisa 2. ENSV Riikliku Hariduskomitee kolleegiumi otsus* (*Appendix 2. Decision of the Council of the State Education Committee of the ESSR*), No. 8–5, June 22, 1989.
13. ESSR, *ENSV Riikliku Hariduskomitee ja ENSV Kehakultuuri ja Spordikomitee käskkiri* (*Decree of the State Education Committee of the ESSR and the Committee for Sports of the ESSR*), No. 286–r287, June 29, 1989.
14. ESSR, *Eesti NSV Haridusministeeriumi käskkiri* (*Decree of the Ministry of Education of the ESSR*), No. 156, June 6, 1990.
15. ER, *Eesti Vabariigi rakendusliku kõrgkooli ajutine põhimäärus, Lisa 1 EV Haridusministeeriumi määrusele* (*Temporary Regulation Concerning Vocational Higher Education Institutions of the Republic of Estonia, Appendix 1 to the Decree of the Ministry of Education of the RE*), No. 4, December 19, 1991.
16. Watson, *Legal Transplants*.
17. ER, Tsiviilseadustiku üldosa seadus (The Law of the General Part of the Civil Code), *Riigi Teataja*, 1994, 53, pp. 889.
18. ER, Erakooliseadus (The Law on Private Schools), *Riigi Teataja*, 1993, pp. 35, 547, revisions: 1995, pp. 12, 119; 1996, pp. 49, 953, 51, 965.
19. ER, Erakooliseadus (The Law of Private Schools), *Riigi Teataja*, 1998, pp. 57, 859.
20. ER, Rakenduskõrgkooli seadus (The Law on Vocational Higher Education Institutions), *Riigi Teataja*, 1998, pp. 61, 980.
21. ER, Ülikooliseadus (The University Law), *Riigi Teataja*, 1995, pp. 12, 119, revisions: 1996, pp. 49, 953; 1996, pp. 51, 965; 1997, pp. 42, 678; 1999, pp. 10, 150.
22. ER, Tartu Ülikooli seadus (The Law of the University of Tartu), *Riigi Teataja*, 1995, pp. 23, 333.
23. ER, Teadus—ja arendustegevuse korralduse seadus (The Law of the Organizing the Research and Development Activities), *Riigi Teataja*, 1997, pp. 30, 471.
24. V. Tomusk, "The Syndrome of the Holy Degree: Critical Reflections on the Staff Development in Estonian Universities," *Higher Education Management*, VII, 3 (1995), pp. 385–397.

25. Tomusk, "Nobody Can Better Destroy Your Higher Education."
26. ER, Kõrghariduse hindamise nõukogu põhikiri (Statute of the Higher Education Evaluation Council), *Kultuuri- ja Haridusministeerium Teataja* (Information Bulletin of the Ministry of Culture and Education), 1995.
27. V. Tomusk, "External Quality Assurance in Estonian Higher Education: Its Glory, Take-Off and Crash," *Quality in Higher Education*, 3, 2(1996), pp. 173–181.
28. V. Tomusk, "When West Meets East: Decontextualizing the Quality of East European Higher Education," *Quality in Higher Education*, 6, 3 (2000), pp. 175–185.
29. Ibid.
30. Tomusk, "Nobody Can Better Destroy Your Higher Education."
31. Tomusk, "Recent Trends in Estonian Higher Education."
32. K. Kroos, *Why There is So Little Reform in Estonian Higher Education: Dual Identity of Estonian Intellectuals*, unpublished M.A. thesis, Budapest: Central European University, 2000.
33. ER, Kõrgharidusreform aastatel 2001–2002 (Higher Education Reform in the years 2001–2002), *Eesti Vabariigi Valitsuse kaskkiri*, June 12, 2001 (*Decree of Government of the Republic of Estonia*, June 12, 2001).
34. As described in D. Levy, *Higher Education and the State in Latin America: Private Challenges to Public Dominance*, Chicago: The University of Chicago Press, 1986.
35. ER, Kõrgharidusreform aastatel 2001–2002.
36. Kroos, "Why there is so little."
37. Ibid.
38. The issue of outsourcing higher education policy making in Eastern Europe is more substantially discussed in the paper: V. Tomusk, *Ministries of Education and Higher Education Policies in Eastern Europe—Steering from Where?* UNESCO Document IIEP/SEM 199. Paris: UNESCO, 2001, presented at the Policy Forum of the Organization of Ministries of Education, IIEP, Paris, June 20–21, 2001.
39. *The European Higher Education Area: Joint Declaration of the European Ministers of Education*, Convened in Bologna on the June 19, 1999.

Chapter 4   Russian Higher Education After Communism: The Candy Man's Gone

1. The initial version of this chapter appeared as "Developments in Russian Higher Education: Legislative and Policy Reform Within a Central and East European Context," *Minerva*, 36, 2 (1998), pp. 125–146, reprinted with the kind permission of the Kluwer Academic Publishers.
2. Quoted in A.K. Sokolov and V.S. Tiazhel'nikova, *Kurs Sovetskoi Istorii 1941–1991* (*The Course of the Soviet History 1941–1991*), Moskva: Vysshaya Shkola, 1999, p. 145. This and all other excerpts from Russian sources quoted in this book have been translated into English by the author.
3. J. Baudrillard, *The Perfect Crime*, London and New York: Verso, 1996.
4. See e.g. V.A. Sadovnichi (Red.), *Obrazovanie kotoroe my mozhem poteryat' (Education that We May Lose)*, Moskva: Moskovskii gosudarstvennyi universitet im. M.V. Lomonosova, 2002.
5. L.R. Graham, *Science, Philosophy, and Human Behavior in the Soviet Union*, New York: Columbia University Press, 1987.
6. "Moldova: Scarred For Life," *Newsweek*, July 16, 2001.
7. Sokolov, *Kurs Sovetskoi Istorii*, p. 249.
8. OECD, *Tertiary Education and Research in the Russian Federation*, Paris: OECD, 1999.
9. Zh.I. Al'ferov and V.A. Sadovnichi, "Obrazovanie dlya Rossii XXI veka (Education for Russia in the 21st Century)," in V.A. Sadovnichi (Red.), *Obrazovanie kotoroe my mozhem poteryat' (Education that We May Lose)*.

10. A.P. Egorshin, "Perspektivy razvitiia obrazovaniia Rossii v XXI v. (Perspectives on the Development of Education in Russia in the 21st century)," *Universitetskoe Upravlenie: Praktika i Analiz (University Management: Practice and Analysis)*, 4 (2000), pp. 50–64.
11. *Tertiary Education and Research in the Russian Federation*.
12. RF, Zakon ob obrazowanii Rossijskoi Federacii (The Law on Education of the Russian Federation), *Rossijskaya Cazeta*, January 26, 1996.
13. V.I. Lenin, Letter to Maxim Gorky, September 15, 1919, in D. Koenker and R. Bachman (Eds.), *Revelations from the Russian Archives* (Washington, DC: Library of Congress, 1997), pp. 229–230.
14. M. David-Fox, *Revolution of the Mind: Higher Learning Among the Bolsheviks, 1918–1929*, Ithaca and London: Cornell University Press, 1997, p. 78.
15. M. Matthews, *Education in the Soviet Union: Politics and Institutions Since Stalin*, London: George Allen & Unwin, 1982.
16. Statement by C.R. Hansen, in *United States–Soviet Relation: 1988, Hearings Before the Subcommittee on the Europe and Middle East of the Committee on Foreign Affairs, House of Representatives, One Hundredth Congress*, vol. I, Washington: U.S. Government Printing Office, 1988, p. 432.
17. *Kurs Sovetskoi Istorii*, p. 146.
18. M. Buttgereit "Higher Education and Its Relations to Employment in the USSR and in the Federal Republic of Germany: A Comparison," in R. Avakov, M. Buttgereit, B.C. Sanyal, and U. Teichler (Eds.), *Higher Education and Employment in the USSR and in the Federal Republic of Germany* (Paris: IIEP, 1984), pp. 231–326.
19. See, e.g. *Review of Higher Education in the Czech and Slovak Federal Republic: Examiner's Report and Questions*, OECD document DEEI.SA/ED/WD (92) 5 (Paris: OECD, 1992).
20. Egorshin, Perspektivy razvitiia obrazovaniia Rossii.
21. Zakon ob obrazowanii, Rossijskoi Federacij.
22. *Tertiary Education and Research in the Russian Federation*.
23. For Feyerabend's view on a philosophy as a profession see P. Feyerabend, *Three Dialogues on Knowledge*, Oxford, UK and Cambridge, USA: Blackwell, 1991.
24. For Khrushchev's 1958 reform, see Matthews, *Education in the Soviet Union*.
25. Hansen, in *United States–Soviet Relations: 1988*, p. 435.
26. Egorshin, Perspektivy razvitiia obrazovaniia Rossii.
27. N. Kovaleva, "Women and Engineering Training in Russia," *European Journal of Education*, 34, 4 (1999), pp. 425–435.
28. St. Petersburg State Technical University, *Koncepcii, Struktury i Soderzhanrie Mnogourovnei Sisremy Vysshego Technicheskogo Obrazovaniia Rossii (Concepts, Structures and Content of Multilevel System of Higher Technical Education a Russia)*, St. Petersburg: State Technical University, 1993.
29. V. Kinelyev, *Preface to State Educational Standard of Higher Professional Education*, Moscow: Goskomvuz, 1995.
30. V. Tomusk, "Discovering the Terra Incognita: Changing Legal Landscape for Higher Education in Estonia," in *The World on the Move and Higher Education in Transition*, Prague: Center for Higher Education Studies, 1995, pp. 147–154.
31. RF, Grazhdanskii Kodeks Rossiiskoi Federacii. Oficial'nyi Tekst (The Civil Code of the Russian Federation. Official Text), Moskva: Ekzamen, 2001.
32. On developments in Estonia, see V. Tomusk, "Recent Trends in Estonian Higher Education: Emergence of the Binary Division from the Point of View of Staff Development," *Minerva*, 34, 3 (Autumn 1996), pp. 279–289.
33. See, e.g. *Croatia: Report of the Advisory Mission on Quality Assurance, Zagreb, 1 S-17 March, 1995*, Council of Europe Document DECS LRP 95/07, Strasbourg: Council of Europe, 1995.
34. David-Fox, *Revolution of the Mind*.

35. RF, *Gosudarstvennyi Obrazovatelnyi Standart Vysshego Professionalnogo Obrazovaniia: Gosudarstvenniie Trebovaniia k minimumu soderzhaniia i urovniu podgotovki vypustnika po specialnosti 071900: Informatsionniie Sistemy v Ekonomike (State Educational Standard for Minimum Content and Level of Training of Graduates from the Programme 071900: Informational Systems in Economics)*, Moscow: Goskomwz, 1995.
36. J. Tayler, "Russia is Finished. The Unstoppable Descent of a Once Great Power into Social Catastrophe and Strategic Irrelevance," *Atlantic Monthly*, May 2001.
37. Egorshin, *Perspektivy razvitiia obrazovaniia Rossii*.
38. D. Piskunov, "Russia: Higher Education and Change," in A.D. Tillett and B. Lesser (Eds.), *An Uncertain Transition: Preliminary Assessment of Higher Education, Science and Technology in Central and Eastern Europe*, Halifax: Dalhousie University, Lester Pearson Institute, 1993.
39. MVShSEN *Obrazovatel'naya politika i obrazovatel'noe sakonodatel'stvo v sovremennoi Rossii. Statisticheskie dannye (Educational Policy and Educational Legislation in Contemporary Russia. Statistical Data)*, Moskva: MVShSEN, 2002.
40. Al'ferov and Sadovnichi, *Obrazovanie dlya Rossii*.
41. L.E. Petrova, *Novye bednye uchenye: zhisnennye strategii v usloviyakh krizisa (New Poor Scholars: Life Strategies Under the Conditions of a Crisis)*, paper presented at the conference Russian Social Sciences: A New Perspective, Moscow, October 7–9, 1999.
42. Ibid.
43. J.P. Naude, "La science russe survit tant bien que mal d'expédients," *Le Monde*, October 16, 1996.
44. *Tertiary Education and Research in the Russian Federation*.
45. Petrova, *Novye bednye uchenye*.
46. Ibid.
47. Ibid.
48. RF, *Natsional'naia doktrina obrazovaniia v Rossiiskoi Federatsii: proekt (National Doctrine of Education of the Russian Fedration: Draft)*, document drafted by the Science and Education Committee of the State Duma of the Russian Federation, 1999.
49. Kinelyev, *State Educational Standard*.
50. V. Tomusk, "Between Politics and Professionalism: Reforming Fundamentals of Higher Education in Central and East Europe," in T. Thanasuthipitak and S.L. Rieb (Eds.), *A Blueprint for Better Graduate Studies*, Chiang Mai: Graduate School, Chiang Mai University, 1997, pp. 331–343.
51. On the concept of the "pure socialist output" in higher education, see L. Balcerowicz, "Research and Education in the Post-Communist Transition," in *Western Paradigms and Eastern Agenda: A Reassessment*, Vienna: Institute of Human Sciences, 1995.
52. On the social sciences in the Soviet Union, see Matthews, *Education in the Soviet Union*.

Chapter 5 Market as Metaphor in East European Higher Education

1. This is a revised version of the article "Market as a Metaphor in Central And East European Higher Education," *International Studies in Sociology of Education*, 8, 2 (1998), pp. 223–240. Reprinted with the kind permission of the Triangle Journals Ltd.
2. G. Neave, "On Living in Interesting Times: Higher Education in Western Europe 1985–1995," *European Journal of Education*, 30 (1995), pp. 377–393.
3. B.R. Clark, *The Higher Education System: Academic Organization in Cross-National Perspective*, Berkeley: University of California Press, 1983.
4. COE, *Regulation of Private Higher Education: Report of the Multilateral Workshop, Prague, 9–11 May 1994*, Council of Europe Document DECS LRP (94) p. 33. Strasbourg: Council of Europe.

5. V. Tomusk, "Between Politics and Professionalism: Reforming Fundamentals of Higher Education in East Europe," in T. Thanasuthopitak and S.L. Rieb (Eds.), *A Blueprint for Better Graduate Studies*, Chiang Mai: Graduate School, Chiang Mai University, 1997, pp. 331–343.
6. Romania, *Carte Blanche of the Reform of Education in Romania*, Bucharest: Ministry of Education, 1995.
7. V. Tomusk, "External Quality Assurance in Estonian Higher Education: Its Glory, Take-Off and Crash," *Quality in Higher Education*, 3 (1997), pp. 173–181.
8. CoE, *Regulation of Private Higher Education: Report of the Multilateral Workshop, Prague, 9–11 May 1994*, Council of Europe Document DECS LRP (94), 33. Strasbourg: Council of Europe.
9. V. Tomusk, "Estonia: Higher Education System," in G. Neave et al. (Eds.), *Education: The Complete Encyclopedia* (CD-ROM), London: Elsevier Science, 1998.
10. T.E. Galko, *Belorusskii kommercheskii universitet upravlenia* (The Belarus State Commercial University of Management), unpublished report, Minsk: BCUM, 1997.
11. CoE, *Croatia: Report of the Advisory Mission on Quality Assurance, Zagreb, 15–17 March 1995*, Council of Europe Document DECS LRP 95/07. Strasbourg: Council of Europe.
12. CoE, *Report of the Advisory Mission to Slovenia: The Draft Law on Organisation and Financing in the Field of Science and Technology, 21–24 February 1996*, Council of Europe Document DECS LRP 96/07. Strasbourg: Council of Europe.
13. CoE, *Estonia: Report of the Advisory Mission, 18–21 May 1994*, Council of Europe Document DECS LRP 94/11. Strasbourg: Council of Europe.
14. D. Levy, *Public Policy for Hungarian Private Higher Education*, unpublished report, 1997.
15. V. Tomusk, "Conflict and Interaction in East European Higher Education: The Triangle of Red Giants, White Dwarfs and Black Holes," *Tertiary Education and Management*, 3 (1997), pp. 247–255.
16. P. Darvas, "Institutional Innovation in East European Higher Education," in M. Szymonski and I. Guzik, (Eds.), *Research at East European Universities*, Krakow: Jagellonian University Press, 1997, pp. 129–149.
17. CoE, *Universities, Colleges and Others: Diversity of Structures for Higher Education: Report of the Multilateral Workshop, Bucharest, 23–25 September 1993*, Council of Europe Document DECS LRP (94) 05. Strasbourg: Council of Europe.
18. R.K. Merton, *Social Theory and Social Structure*, New York: Free Press, 1968.
19. Editorial, *Nasha Gazeta*, November 15, 1997.
20. KAF, *1997 Yearbook of the Kyrgyz American Faculty*, Bishkek: Kyrgyz-American School, Kyrgyz State National University, 1997.
21. Levy, *Public Policy*.
22. Tomusk, "Conflict and Interaction," pp. 247–255.
23. CoE, *Brain Drain from Universities: Final Report of the CC-PU Forum Role Conference*, Council of Europe Document DECS-HE 94/26. Strasbourg: Council of Europe.
24. Levy, *Public Policy*.
25. V. Tomusk, "Recent Trends in Estonian Higher Education: Emergence of the Binary Division from the Point of View of Staff Development," *Minerva*, 34 (1996), pp. 279–289.
26. O. Kivinen, and R. Rinne, "State, Governmentality and Education—the Nordic experience," *British Journal of Sociology of Education*, 19 (1998), pp. 39–52.
27. See, e.g. S. Slantcheva, "The Challenges of Vertical Degree Differentiation Within Bulgarian Universities: The Problematic Introduction of the Three-Level System of Higher Education," *Tertiary Education and Management*, 6, 3 (2000), pp. 209–225.
28. V. Tomusk, "Developments in Russian Higher Education: Legislative and Policy Reforms Within East European Context," *Minerva*, 36, 3 (1998), pp. 125–146.
29. V. Tomusk, "Quality in Transition: Attributing a Meaning to Quality in Central-East European Higher Education," in J.L. Lambert and W. Banta (Compilers), *Proceedings of*

the *Eighth International Conference on* Assessing *Quality in Higher Education, 14–16 July, 1996, Queensland, Australia,* Indianapolis: Indiana University-Purdue Indianapolis, 1996, pp. 291–301.
30. Levy, *Public Policy*.
31. Plato, "Socrates Defence (Apology)," in E. Hamilton and H. Cairns (Eds.), *Plato: The Collected Dialogues*, Princeton: Princeton University Press, 1961.
32. Tomusk, "Developments in Russian Higher Education," pp. 125–146.
33. S.P. Heyneman, "The Transition from Parry State to Open Democracy: The Role of Education," *International Journal of Educational Development*, 2000.
34. Neave, "On Living in Interesting Times," pp. 377–393.
35. Tomusk, "Developments in Russian Higher Education," pp. 125–146.
36. V. Tomusk, "The Syndrome of the Holy Degree: Critical Reflections on the Staff Development in Estonian Universities," *Higher Education Management*, 7, 3 (1995), pp. 385–397.
37. P. Bourdieu, J.C. Passeron, and M. de Saint Martin, *Academic Discourse*, Cambridge: Polity Press, 1994.
38. Tomusk "Conflict and Interaction," pp. 247–255.
39. M.W. Apple, *Cultural Politics and Education*, London: Open University Press, 1996.
40. OECD, *Review of Higher Education in the Czech and Slovak Federal Republic: Examiner's Report and Questions*, OECD Document DEELSA/ED/WD (92) 5. Paris: OECD, 1992.
41. *Croatia: Report of the Advisory Mission on Quality Assurance, Zagreb, 15–17 March 1995.*
42. L. Maior, and L. Georgescu, *Status of the Higher Education Reform in Romania*, unpublished report for the World Bank, 1996.

Chapter 6   Reaching Beyond Geometry—The Privateness of
Private Universities

1. A part of this chapter appeared in the article by V. Tomusk, "Conflict and Interaction in Central and East European Higher Education: The Triangle of Red Giants, White Dwarfs and Black Holes," *Tertiary Education and Management*, 3, 3 (1997), pp. 249–257, reprinted with the kind permission of Kluwer Academic Publishers.
2. J. Baudrillard, *The Illusion of the End*, Cambridge: Polity Press, 1994.
3. R. Dahrendorf, *Universities After Communism: The Hannah Arendt Prize and the Reform of Higher Education in East Central Europe*, Hamburg: Edition Körber-Stiftung, 2000.
4. Baudrillard, *The Illusion of the End*.
5. B.R. Clark, *The Higher Education System: Academic Organization in Cross-National Perspective*, University of California Press, 1983.
6. A. Watson, *Legal Transplants: An Approach to Comparative Law*, Second edition, Athens and London: The University of Georgia Press, 1993.
7. M. Matthews, *Education in the Soviet Union: Policies and Institutions Since Stalin*, London: George Allen and Unwin, 1982.
8. V. Tomusk, "Recent Developments in Estonian Higher Education: Emergence of the Binary Division from the Point of View of Staff Development," *Minerva*, 34, 3 (1996), pp. 279–289.
9. P. Darvas, *Institutional Innovations in Higher Education in Central Europe: Agenda for Research and Cooperation*, Vienna: Institute for Human Sciences, 1996.
10. J. Baudrillard, *Simulacres et Simulation*, Paris: Galilée, 1981.
11. H. Winkler, "Funding of Higher Education in Germany," in P. Hare (Ed.), *Structure and Financing of Higher Education in Russia, Ukraine and the EU*, London: Jessica Kingsley Publishers, 1997.
12. D. Levy, *Higher Education and the State in Latin America: Private Challenges to Public Dominance*, Chicago: The University of Chicago Press, 1986.

13. T. Leary, *Interpersonal Diagnosis of Personality*, New York: John Wiley & Sons, 1957.
14. Information given by Mrs. Tatyana Galko, former Deputy Minister of Education, in Minsk, May 1998.
15. Levy, *Higher Education and the State in Latin America*.
16. EHU, *Uchreditel'nyi dogovor o sozdanii i deyatel'nosti evropejskogo gumanitarnogo universiteta* (Founding Agreement on the Establishment and Activities of the European Humanities University), unpublished document, Minsk: EHU, 1997.
17. Levy, *Higher Education and the State in Latin America*.
18. Ibid.
19. IHF, *International Humanitarian Foundation*, Minsk: IHF, 1998.
20. Levy, *Higher Education and the State in Latin America*.
21. IHF, *International Humanitarian Foundation*.
22. EHU, *Ustav Evropeiskogo Gumanitarnogo Universiteta (Charter of the European Humanities University)*, unpublished document, Minsk: EHU, 1997.
23. Dahrendorf, *Universities After Communism*, p. 74.
24. Levy, *Higher Education and the State in Latin America*.
25. Ibid.
26. S. Pavlychko, et al., *European Humanities University: Report of the Review Panel*, unpublished report of HESP/OSI, 1998.
27. EHU, *The Current State and Prospects for the Development of the European Humanities University*, unpublished report, Minsk: EHU, 1998.
28. Leary, *Interpersonal Diagnosis of Personality*.
29. Studies for the first research degree, sometimes considered as Ph.D. equivalent.
30. Higher doctorate, equivalent to Dr. Habil.

Chapter 7 Exploring the Limits of Entrepreneurial Response

1. This chapter is a revised version of the paper *Exploring the Limits of the Entrepreneurial Response: Academia Istropolitana Nova*, presented at the 22nd Annual EAIR Forum, Freie Universität Berlin, Germany, September 6–9, 2000.
2. First Book of Moses, Genesis 3:19; English translation from the Authorised King James Version.
3. B.R. Clark, *Creating Entrepreneurial Universities: Organizational Pathways of Transformation*, London: IAU Press/Elsevier Science Ltd., 1998.
4. B. Readings, *University in Ruins*, Cambridge, Massachusetts, and London, England: Harvard University Press, 1996.
5. G. Neave, "On Living in Interesting Times: Higher Education in Western Europe 1985–1995," *European Journal of Education*, 30 (1995), pp. 377–393.
6. E. Gellner, *Nationalism*, London: Phoenix, 1998.
7. P. Bourdieu, *The State Nobility*, Stanford: Stanford University Press, 1996.
8. M. Skillbeck, *The Challenge of Diversifying Tertiary Education: OECD Responses*, paper presented at the World Bank/Open Society Institute Workshop on Diversification of Higher Education in Central and Eastern Europe, Budapest, November 18–20, 1999.
9. M. Anderson, *Impostors in the Temple: A Blueprint for Improving Higher Education in America*, Stanford: Hoover Institution Press, 1996.
10. C. Colbeck, "Reshaping Forces that Perpetuate the Research/practice Gap: Focus on New Faculty," in A. Kezar and P. Eckel (Eds.), *Higher Education Research: Transfer to Practice. New Directions for Higher Education*, San Francisco: Jossey-Bass, 2000.
11. G. Neave and F. VanVught, *Prometheus Bound: The Changing Relationship Between Government and Higher Education in Western Europe*, Oxford, New York, Beijing, Frankfurt, São Paulo, Sydney, Tokyo, Toronto: Pergamon Press, 1991.
12. J.L. Davies, *The Entrepreneurial and Adaptive University*, Paris: OECD, 1987.

13. Readings, *University in Ruins*.
14. Ibid.
15. Information shared by Ivan Ostrovski at a Regional TEMPUS Seminar in the Framework of the Stability Pact for South East Europe, *The Link Between Strategic University Management and Higher Education Policy*, Sarajevo, June 15–18, 2000.
16. V. Savchuk, P. Luzik, I. Gal, and V. Oparin, "Higher Education in Ukraine: Structure and Financing," in P. Hare (Ed.), *Structure and Financing of Higher Education in Russia, Ukraine and the EU*, London: Jessica Kingsley Publishers, 1997.
17. R. Dahrendorf, *Universities After Communism: The Hannah Arendt Prize and the Reform of Higher Education in East Central Europe*, Hamburg: Edition Körber-Stiftung, 2000.
18. AI, *Report 1991–1995: The Hannah Arendt Prize 1996 Self-Study Questionnaire*, unpublished document, Bratislava: Academia Istropolitana, 1996.
19. Ibid.
20. R.L. Geiger, *Single Donor Universities*, PONP Working Paper No. 215 and ISPS Working Paper No. 2215, Yale University, Institution for Social and Policy Studies, 1995.
21. P. Darvas, *Institutional Innovation in Central European Higher Education*, Vienna: Institute of Human Sciences, 1996.
22. *Report 1991–1995*.
23. Darvas, *Institutional Innovation*.
24. AIN, *AINova—The Vision for Future*, unpublished document, Svjaty Jur: Academia Istropolitana Nova,1999.
25. Ibid.
26. B. Bollag, Private Colleges Reshape Higher Education in Eastern Europe and Former Soviet States, *The Chronicle of Higher Education*, June 11, 1999.
27. *Future Prospectus: Academia Istropolitana Nova*, unpublished document, Svjaty Jur: Academia Istropolitana Nova, 1998.

Chapter 8   When East Meets West: Decontextualizing the Quality of East European Higher Education

1. This is a revised version of an article that appeared under the same title in *Quality in Higher Education*, 6, 3 (2000), pp. 175–185, reprinted with the kind permission of Taylor & Francis Ltd.
2. Lao-tzu, *Te-Tao Ching*, New York: The Modern Library, 1993.
3. V. Tomusk, "Quality in Transition: Attributing a Meaning to Quality in Central-East European Higher Education," in Jane L. Lambert and Trudy W. Banta (Compilers), *Proceeding of the Eighth International Conference on Assessing Quality in Higher Education, July 14–16, 1996, Queensland, Australia*, pp. 291–301.
4. J.L. Ratcliff, "Institutional Self-Evaluation and Quality Assurance: A Global View," in A. Strydom and L. Lategan (Eds.), *Institutional Self-Evaluation in Higher Education in South Africa*, Blomfontein, RSA: Unit for Higher Education Research, University of the Orange Free State, 1998.
5. P. Sterian, *Quality Assurance System in Romanian Higher Education*, paper presented at the Regional Training Seminar for Quality Assurance in Higher Education: Self Assessment and Peer Review, Budapest, November 10–16, 1996.
6. Ratcliff, "Institutional Self-Evaluation and Quality Assurance."
7. E.A. Jones and J.L. Ratcliff, "Global Perspectives on Program Assessment and Accreditation," in L. Lategan, M. Fourie, and A. Strydom (Eds.), *Programme Assessment in Higher Education in South Africa*, Blomfontein, RSA: Unit for Higher Education Research, University of the Orange Free State, 1999.
8. M.M.H. Frederiks, D.F. Westerheijden, and P.J.M. Weusthof, "Effects of Quality Assessment in Dutch Higher Education," *European Journal of Education*, 29, 2 (1994), pp. 181–199.

9. V. Tomusk, "Conflict and Interaction in Central and East European Higher Education: The Triangle of Red Giants, White Dwarfs and Black Holes," *Tertiary Education and Management*, 3, 3 (1997), pp. 247–255.
10. P. Broadfoot, "Quality Standards and Control in Higher Education: What Price Life-Long Learning?" *International Journal of Sociology of Education*, 8, 2 (1998), pp. 155–180.
11. V. Tomusk, "Developments in Russian Higher Education: Legislative and Policy Reform Within Central and East European Context," *Minerva*, 36, 2 (1998), pp. 125–146.
12. E. Gilder, *Report on the UNESCO/CEPES Seminar "Quality Management in Higher Education," Bucharest, May 23, 1996*, unpublished document.
13. G. Johnes, "The Funding of Higher Education in the United Kingdom," in P. Hare (Ed.), *Structure and Financing of Higher Education in Russia, Ukraine and the EU*, London: Jessica Kingsley Publishers, 1997.
14. J.A. Getty, and O.V. Naumov, *The Road to Terror: Stalin and the Self-Destruction of the Bolsheviks, 1932–1939*, New Haven and London: Yale University Press, 1999, p. 388.
15. V. Tomusk, "External Quality Assurance in Estonian Higher Education: Its glory, Take-Off and Crash," *Quality in Higher Education*, 3, 2 (1997), pp. 173–181.
16. Tomusk, "Quality in Transition," pp. 291–301.
17. Tomusk, "Developments in Russian Higher Education," pp. 125–146.
18. See e.g. V. Savchuk, P. Luzik, I. Gal, and V. Oparin, "Higher Education in Ukraine: Structure and Financing," in P. Hare (Ed.), *Structure and Financing of Higher Education in Russia, Ukraine and the EU*, London: Jessica Kingsley Publishers, 1997.
19. Ratcliff, "Institutional Self-Evaluation and Quality Assurance."
20. Ibid.
21. J. Brennan, L.C.J. Goedegebuure, D.F. Westerheijden, and T. Shah, *Comparing Quality in Europe*, publication no. 101 in the series Higher Education Policy Studies, Enschede: CHEPS, University of Twente, 1991.
22. R.M. Pirsig, *Zen and the Art of the Motorcycle Maintenance: An Inquiry into Values*, London: Vintage, 1974.
23. H. Putnam, *Words and Life*, Cambridge MA: Harvard University Press, 1994.
24. RF, *Podgotovka specialistov v oblasti gumanitarnyk i social'no-ekonomicheskik nauk* (Preparation of Experts in Humanities and Social Sciences), Moskva: Goskomvuz, 1995.
25. For detailed description of the standard see Tomusk, "Developments in Russian Higher Education," pp. 125–146.
26. E. Kovaleva, *Progress and Issues in Reforming Social Science Teaching in Ukraine*, paper presented at the Second Annual CEP Eastern Scholars Roundtable, Lviv State University, Lviv, Ukraine, May 1999.
27. S. Baron, S. Riddell, and A. Wilson, "The Secret of Eternal Youth: Identity, Risk and Learning Difficulties," *British Journal of Sociology of Education*, 20, 4 (1999), pp. 483–499.
28. J. Aaviksoo, "Mõtteid ülikooli arengust (Thoughts on Development of the University)," *Tartu Ülikool*, 3, 4 (1992), pp. 39–44.
29. See *Natsional'naia doktrina obrazovaniia v Rossijskoi Federatsii: proekt* (*National Doctrine of Education of the Russian Fedration: Draft*), document drafted by the Science and Education Committee of the State Duma of the Russian Federation, 1999.
30. L. Cerych, "Educational Reforms in Central and Eastern Europe," *European Journal of Education*, 30, 4 (1995), pp. 423–435.

Chapter 9   Reproduction of the "State Nobility" in Eastern Europe: Past Patterns and New Practices

1. This is a revised version of an article that appeared under the same title in the *British Journal of Sociology of Education*, 21, 2 (2000), pp. 269–283, reprinted with the kind permission of Taylor & Francis Ltd.

2. Mao Tse-Tung, *Quotations from Chairman Mao Tse Tung*, San Francisco: China Books and Periodicals, 1990.
3. P. Bourdieu, *The State Nobility*, Stanford: Stanford University Press, 1996.
4. I.B. Neumann, *Uses of the Other: "The East" in European Identity Formation*, Manchester: Manchester University Press, 1999.
5. V. Tomusk, "Conflict and Interaction in Central and East European Higher Education: The Triangle of Red Giants, White Dwarfs and Black Holes," *Tertiary Education and Management*, 3, 3 (1997), pp. 247–255.
6. Bourdieu, *The State Nobility*.
7. L.J.D. Wacquant, foreword in Bourdieu, *State Nobility*.
8. P. Drucker, "Defending Japanese Bureaucrats," *Foreign Affairs*, September/October 1998.
9. See e.g. M. Anderson, *Impostors in the Temple: A Blueprint for Improving Higher Education in America*, Stanford: Hoover Institution Press, 1996.
10. Bourdien, *The State Nobility*.
11. Ibid.
12. D. Levy, *Higher Education and the State in Latin America: Private Challenges to Public Dominance*, Chicago: The University of Chicago Press, 1986.
13. G. Neave, "On Living in Interesting Times: Higher Education in Western Europe 1985–1995," *European Journal of Education*, 30, 4 (1995), pp. 377–393.
14. V.I. Lenin, letter to Maxim Gorky, September 15, 1919, in D. Koenker and R. Bachman, (Eds.), *Revelations from the Russian Archives*, Washington, DC: Library of Congress, 1997.
15. I. Berlin, "The Silence in Russian Culture," *Foreign Affairs*, October 1957.
16. L.R. Graham, *Science, Philosophy, and Human Behavior in the Soviet Union*, New York: Columbia University Press, 1987.
17. C. Offe, *Modernity and the State: East, West*, Cambridge, Massachusetts, and London, England: The MIT Press, 1996.
18. A. Wellmer, *Endgames: The Irreconcilable Nature of Modernity*, Cambridge: Massachusetts, and London, England: The MIT Press, 1998.
19. M.S. Gorbachev, *Perestroika i novoe myshlenie dlia nashei strany i dlia vsego mira* (Perestroika and New Thinking for Our Country and for the Whole World), Moskva: Politizdat, 1987.
20. E. Todd, *The Final Fall: An Essay on the Decomposition of the Soviet Sphere*, New York: Karz Publishers, 1979.
21. N. Chomsky, *The Old and the New Cold War*, New York: Pantheon Books, 1980.
22. Berlin, "The Silence in Russian Culture."
23. L.R. Graham, *Science, Philosophy, and Human Behavior in the Soviet Union*, New York: Columbia University Press, 1987.
24. V.P. Mokhov, "Stratifikaciia sovetskoi regional'noi politicheskoi elity. 1960–1990 gg. (Stratification of the Soviet Regional Political Élite, 1960s–1990s)," in M.N. Afanas'ev (Ed.), *Vlast' i obshchestvo v postsovetskoi Rossii: novye praktiki i instituty* (Power and Society in Post-Soviet Russia: New Practices and Institutions), Moscow: Moscow Public Science Foundation, 1999.
25. A. Giddens, *The Third Way: The Renewal of Social Democracy*, Cambridge, Polity Press, 1998.
26. B.R. Clark, *Academic Power in Italy: Bureaucracy and Oligarchy in a National University System*, Chicago and London: The University of Chicago Press, 1977.
27. U. Teichler, *Changing Patterns of the Higher Education System: Experience of Three Decades*, London: Jessica Kingsley Publishers, 1988.
28. D. Chuprunov, R. Avakov, and E. Jiltsov, "Higher Education, Employment and Technological Progress in the USSR," in R. Avakov, M. Buttgereit, B.C. Sanyal, and U. Teichler (Eds.), *Higher Education and Employment in the USSR and in the Federal Republic of Germany*, Paris: UNESCO/IIEP, 1984.

29. C. Gaddy and B. Ickes, "Russia's Virtual Economy," *Foreign Affairs*, September/October 1998.
30. M. Matthews, *Education in the Soviet Union: Politics and Institutions Since Stalin*, London: George Allen & Unwin, 1982.
31. Giddens, *The Third Way*.
32. M. Matthews, *Patterns of Deprivation in the Soviet Union Under Brezhnev and Gorbachev*, Stanford: Hoover Institution Press, 1989.
33. Mokhov, "Stratifikaciia sovetskoi regional'noi politicheskoi elity."
34. S.D. Roper, *Romania: The Unfinished Revolution*, Amsterdam: Harwood Academic Publishers, 2000, p. 51.
35. R. Frydman, K. Murphy, and A. Rapaczynski, *Capitalism with a Comrade's Face*, Budapest: Central European University Press, 1998.
36. A.J. Motyl, "After Empire: Competing Discourses an Inter-State Conflict in Post Imperial Eastern Europe," in B.R. Rubin and J. Snyder (Eds.), *Post-Soviet Political Order*, London: Routledge, 1998.
37. M. Stepko, "Ukraine," in R. in't Veld, H.-P. Füssel, and G. Neave, (Eds.), *Relations Between State and Higher Education*, The Hague: Kluwer Law International, 1996.
38. V. Savchuk, P. Luzik, I. Gal, and V. Oparin, "Higher Education in Ukraine: Structure and Financing," in P. Hare (Ed.), *Structure and Financing of Higher Education in Russia, Ukraine and the EU*, London: Jessica Kingsley Publishers, 1997.
39. P. Darvas, *Institutional Innovation in Central European Higher Education*, Vienna: Institute for Human Sciences, 1996.
40. Bourdien, *The State Nobility*.
41. I. Prifti, "Albania," in Veld, Füssel, and Neave, (Eds.), *Relations Between State and Higher Education*.
42. Veld, Füssel and Neave, (Eds.), *Relations Between State and Higher Education*.
43. N. Chomsky, *Language and Freedom*, New York: Pantheon Books, 1970.

Chapter 10 The Communication Community and the Scam of the Knowledge Society

1. This is a revised version of the paper *The Communication Community under Construction: Is There a Way to Increase Public Accountability of Social Research?* presented at the International Symposium on a Blueprint for Better Graduate Studies—Grad-Blueprint 2000, Graduate School of the Prince Songkla University, Phuket, Thailand, November 22–24, 2000.
2. Quoted in J.A. Murphy, "Degree of Waste: The Economic Benefits of Educational Expansion," *Oxford Review of Education*, 19, 1 (1993), pp. 9–31.
3. M. Godet, *From Anticipation to Action: A Handbook of Strategic Prospective*, Paris: UNESCO Publishing, 1993.
4. A. Hickling Hudson, "The Cuban University and Educational Outreach: Cuba's Contribution to Post-Colonial Development," in B. Teasdale and Z. Ma Rhea (Eds.), *Local Knowledge and Wisdom in Higher Education*, Oxford: Pergamon, 2000.
5. A. Schutz, *On Phenomenology and Social Relations*, Chicago: University of Chicago Press, 1975.
6. Quoted in L.J. Teffo, "Africanist Thinking: An Invitation to Authenticity," in P. Higgs, N.C.G. Vakalisa, T.V. Mda, and N.T. Assié-Lumumba (Eds.), *African Voices in Education*, Lansdowne: Juta, 2000, p. 106.
7. T. Becher, *Academic Tribes and Territories: Intellectual Enquiry and the Cultures of Disciplines*, Buckingham and Bristol: Open University Press, 1989.
8. B.R. Clark, *The Higher Education System: Academic Organization in Cross-National Perspective*, Berkeley and Los Angeles: University of California Press, 1983.

9. OECD, *Education and Learning for Sustainable Consumption*, OECD document COM/ENV/CERI (99) 64, Paris: OECD, 1999.
10. J. Murphy, "A Degree of Waste: The Economic Benefits of Educational Expansion," *Oxford Review of Education*, 19, 1 (1993), pp. 9–31.
11. E. Gellner, *Nationalism*, London: Phoenix, 1998.
12. M. Tannehill and L. Tannehill, *The Market for Liberty*, New York: Libertarian Review Foundation, 1984, p. 155.
13. P. Bourdieu, *The State Nobility*, Stanford: Stanford University Press, 1996.
14. Gellner, *Nationalism*.
15. Bourdieu, *The State Nobility*.
16. N. Matos, "The Nature of Learning, Teaching and Research in Higher Education in Africa," in Higgs, Vakalisa, Mda, and Assié-Lumumba (Eds.), *African Voices in Education*, p. 24.
17. K.R. Popper, *Alles Leben ist Problemlösen: Über Erkenntnis, Geschichte und Politik*, München, Zürich: Piper, 1994.
18. C.A. Odora Hoppers, "African Voices in Education: Retrieving the Past, Engaging the Present and Shaping the Future," in Higgs, Vakalisa, Mda, and Assié-Lumumba (Eds.), *African Voices in Education*, p. 7.
19. H.W. Vilakazi, "The Problem of Education in Africa," in Higgs, Vakalisa, Mda, and Assié-Lumumba (Eds.), *African Voices in Education*, p. 197.
20. C. Offe, "Democracy and Trust," *Theoria: A Journal of Social and Political Theory*, 96 (2000), pp. 1–13.
21. M. McLuhan, *Understanding Media: The Extensions of Man*, Cambridge, Massachusetts and London, England: The MIT Press, 1964/94.
22. K. Bales, *Disposable People: New Slavery in the Global Economy*, Berkeley, Los Angeles, London: University of California Press, 1999.
23. OECD, *Education and Learning for Sustainable Consumption*.
24. P. Sztompka, "Conceptual Frameworks in Comparative Inquiry: Divergent or Convergent?" in M. Albrow and E. King (Eds.), *Globalization, Knowledge and Society*, London, Thousand Oaks, New Delhi: Sage Publications, 1990.
25. B. Readings, *University in Ruins*, Cambridge, Massachusetts, and London, England: Harvard University Press, 1996.
26. Ibid.
27. Ibid.
28. Matos, "The Nature of Learning," p. 13.
29. T. Lumumba-Kasongo, "Rethinking Educational Paradigms in Africa: Imperatives for Social Progress in the New Millennium," in Higgs, Vakalisa, Mda, and Assié-Lumumba (Eds.), *African Voices in Education*, p. 143.
30. P.J. Hountondji, "Manufacturing Unemployment: The Crisis of Education in Africa," in Higgs, Vakalisa, Mda, and Assié-Lumumba (Eds.), *African Voices in Education*, p. 41.
31. World Bank, *Constructing Knowledge Societies. New Challenges for Tertiary Education: A World Bank Strategy*, Washington, DC: The World Bank, Education Group, Human Development Network, 2002.
32. A. Hoogvelt, *Globalization and the Postcolonial World: The New Political Economy of Development*, Second edition, Hampshire: Palgrave, 2001.
33. Lumumba-Kasonga, "Rethinking Educational Paradigms," p. 153.
34. M. Nkomo, "Educational Research in the African Development Context: Rediscovery, Reconstruction and Prospects," in Higgs, Vakalisa, Mda, and Assié-Lumumba (Eds.), *African Voices in Education*, p. 54.
35. Gellner, *Nationalism*.
36. Quoted in Teffo, "Africanist Thinking," p. 108.
37. OECD, *Education and Learning for Sustainable Consumption*.

38. M. Polányi, *Science, Faith and Society*, Chicago and London: University of Chicago Press, 1946.
39. P. Gale, "Indigenous Rights and Higher Education in Australia: 'Not Just Black and White,'" in B. Teasdale and Z. Ma Rhea (Eds.), *Local Knowledge and Wisdom in Higher Education*, Oxford: Pergamon, 2000.
40. Gellner, *Nationalism*.
41. Nkomo, "Educational Research in the African Development Context," p. 48.
42. Ibid., p. 53.
43. Matos, "The Nature of Learning, Teaching and Research," p. 20.
44. Vilakezi, "The Problem of Education in Africa," p. 198.
45. I.N. Goduka, "African/Indigenous Philosophies: Legitimizing Spiritually Centred Wisdoms within the Academy," in Higgs, Vakalisa, Mda, and Assié-Lumumba (Eds.), *African Voices in Education*, p. 80.
46. S. Seepe, "Africanization of Knowledge: Exploring Mathematical and Scientific Knowledge Embedded in African Cultural Practices," in Higgs, Vakalisa, Mda, and Assié-Lumumba (Eds.), *African Voices in Education*, p. 119.
47. Offe, "Democracy and Trust."
48. J.C. Lilly, *The Scientist: A Metaphysical Autobiography*, Berkeley, California: Ronin Publishing, 1996.
49. Nkomo, "Educational Research in the African Development Context," pp. 53–54.
50. B. Brock-Utne, "Transforming African Universities Using Indigenous Perspectives and Local Experience," in B. Teasdale and Z. Ma Rhea (Eds.), *Local Knowledge and Wisdom in Higher Education*, Oxford: Pergamon, 2000.
51. Readings, *University in Ruins*.
52. Seepe, "Africanization of Knowledge," p. 119.
53. See e.g. R. Collins, *The Sociology of Philosophies*, Cambridge, Massachusetts, and London, England: The Belknap Press of Harvard University Press, 1998.
54. R.J. Herrnstein and C. Murray, *The Bell Curve: Intelligence and Class Structure in American Life*, New York, London, Toronto, Sidney, Tokyo, Singapore: Free Press, 1994.
55. Gellner, *Nationalism*.
56. T. Kuhn, *The Structure of Scientific Revolutions*, second edition, enlarged, Chicago and London: University of Chicago Press, 1962.
57. Becher, *Academic Tribes and Territories*.
58. M. Gibbons, C. Limoges, H. Nowotny, S. Schwartzman, P. Scott, and M. Trow, *The New Production of Knowledge: The Dynamics of Science and Research in Contemporary Societies*, London, Thousand Oaks, New Delhi: SAGE Publications, 1994.
59. C. Colbeck, "Reshaping Forces that Perpetuate the Research/Practice Gap: Focus on New Faculty." in A. Kezar and P. Eckel (Eds.), *Higher Education Research: Transfer to Practice. New Directions for Higher Education*, San Francisco: Jossey-Bass, 2000.
60. J. Habermas, *The New Conservatism*, Cambridge: Polity Press, 1989.

Chapter 11 Transnational Capitalist Class and World Bank "Aid" for Higher Education

1. This is a revised version of an article that appeared under the same title in *International Studies in Sociology of Education*, 12, 3 (2002), pp. 337–354. Reprinted with the kind permission of Triange Journals Ltd.
2. V. Hatar, *The Right to Sanity*, Budapest: Corvina Books Ltd., 1999.
3. I. Ramonet, "L'axe du Mal," *Le Monde Diplomatique*, March 2002.
4. L. Sklair, *The Transnational Capitalist Class*, Oxford and Malden: Blackwell Publishers, 2001.
5. M. McLuhan, *War and Peace in the Global Village*, San Francisco: HardWired, 1968/97.

6. Sklair, *The Transnational Capitalist Class*, p. 29.
7. Ibid., p. 2.
8. Ibid., p. 13.
9. Ibid., p. 4.
10. Ibid., pp. 20–21.
11. Ibid., p. 3.
12. Ibid., p. 15.
13. Ibid., p. 17.
14. D. Brooks, *Bobos in Paradise: The New Upper Class and How They Got There*, New York: Simon & Schuster, 2000.
15. A. Kazmin, "Rush to Plug Vietnam's Learning Gap," *The Financial Times*, July 2, 2002.
16. See e.g. J. Lukacs, *At the End of an Age*, New Haven & London: Yale University Press, 2002.
17. M. Henry, B. Lingard, F. Rizvi, and S. Taylor, *The OECD, Globalization and Education Policy*, Amsterdam: Pergamon, 2001.
18. R. Kent, *Two Positions in the International Debate About Higher Education: The World Bank and UNESCO*, paper presented at the 1995 meeting of the Latin American Studies Association: Washington, DC, September 28–30, 1995.
19. Ibid.
20. G. Hancock, *Lords of Poverty: The Power, Prestige, and Corruption of the International Aid Business*, New York: Atlantic Monthly Press, 1989.
21. D.P. Moynihan, "The Professionalization of Reform," *The Public Interest*, 1 (1965), p. 10.
22. Ibid., p. 16.
23. See e.g. H. Marcuse, *One-Dimensional Man: Studies in the Ideology of Advanced Industrial Society*, London: Routledge, 1964/91.
24. Quoted in Z. Bauman, "Pierre Bourdieu or the Dialectics of Vita Contemplativa and Vita Activa," *Revue International de Philosophie*, 56, 2 (2002), pp. 179–193.
25. A. Heller, *The Three Logics of Modernity and the Double Bind of the Modern Imagination*, public lecture series no. 23, Budapest: Collegium Budapest, 2000.
26. *OECD Education and Learning for Sustainable Consumption*, OECD document COM/ENV/CERI (99) 64, Paris: OECD, 1999.
27. E. Gellner, *Conditions of Liberty: Civil Society and Its Rivals*, New York: Allen Lane/Penguin, 1994.
28. Bauman, "Pierre Bourdieu or the Dialectics of Vita Contemplativa and Vita Activa," pp. 179–193.
29. R.M. Solow, "Notes on Social Capital and Economic Performance," in P. Dasgupta and I. Serageldin (Eds.), *Social Capital: A Multifaceted Perspective*, Washington, DC: The World Bank, 2000.
30. J.S. Coleman, "Social Capital in the Creation of Human Capital," in Dasgupta and Serageldin (Eds.), *Social Capital: A Multifaceted Perspective*.
31. J. Salmi, "Globalization and the Knowledge-Based Economy Pose Challenges to all Purveyors of Tertiary Education. Tertiary Education—News Item," *Times Higher Education Supplement*, December 3, 1999.
32. J. Lennon, *The John Lennon Collection*, London: EMI Records, 1989.
33. World Bank, *Constructing Knowledge Societies: New Challenges for Tertiary Education*, Washington, DC: The World Bank, 2002.
34. Ibid., p. 113.
35. World Bank, *Challenges to Education Systems in Transition Economies*, Washington, DC: The World Bank, 2000.
36. J. Einhorn, "The World Bank's Mission Creep," *Foreign Affairs*, September/October 2001, pp. 22–35.
37. Hancock, *Lords of Poverty*.

38. World Bank, *Constructing Knowledge Societies*, p. 104.
39. World Bank, *Staff Appraisal Report: Russian Federation, Education Innovation Project. Report No. 16267-RU*, Washington, DC: The World Bank/Human Resources Operations Division, Country Department III, Europe and Central Asia Region, 1997, p. 39.
40. World Bank, *Staff Appraisal Report: Romania, Reform of Higher Education and Research Project. Report No. 15525-RO*, Washington, DC: The World Bank/Human Resources Operations Division, Country Department I, Europe and Central Asia Region, 1996, p. 35.
41. Coopers and Lybrand, *Hungarian Higher Education: Content and Development of HEI Business Plans. Final Report for the World Bank*, unpublished document, 1997.
42. A.W. Morgan and A.A. Bergerson, "Importing Organizational Reform: The Case of Lay Boards in Hungary," *Higher Education*, 40 (2000), pp. 423–448.
43. Hancock, *Lords of Poverty*.
44. World Bank, *Staff Appraisal Report: Romania, Education Reform Project. Report No. 11931-RO*, Washington, DC: The World Bank/Human Resources Operations Division, Central and Southern Europe Departments, Europe and Central Asia Region, 1994.
45. World Bank, *Staff Appraisal Report*, 1996.
46. World Bank, *Staff Appraisal Report*, 1997.
47. World Bank, *Staff Appraisal Report*, 1994.
48. Hancock, *Lords of Poverty*.
49. World Bank, *Constructing Knowledge Societies*, 2002, p. xxv.
50. World Bank, *Constructing Knowledge Societies: New Challenges for Tertiary Education. A World Bank Strategy*, unpublished draft, Washington, DC: The World Bank, 2001, p. 65.
51. Hancock, *Lords of Poverty*, p. 181.
52. OECD, *Education and Learning for Sustainable Consumption*.

Chapter 12   Toward a Model of Higher Education Reform in Central and East Europe

1. I. Rév, "The-Self-Not-Fulfilling Prophecy," in R. Dahrendorf, Y.Elkana, A. Neier, W. Newton-Smith, and I. Rév (Eds.), *The Paradoxes of Unintended Consequences*, Budapest and New York: Central European University Press, 2000.
2. L. Cerych and P. Sabatier, *Great Expectations and Mixed Performance: The Implementation of Higher Education Reforms in Europe*, Trentham: Trentham Books, 1986.
3. Rév, "The-Self-Not-Fulfilling Prophecy."
4. V. Tomusk, "Reproduction of the 'State Nobility' in Eastern Europe: Past Patterns and New Practices," *British Journal of Sociology of Education*, 21, 2 (2000), pp. 269–283.
5. Rév, "The-Self-Not-Fulfilling Prophecy."
6. E. Kovaleva, *Progress and Issues in Reforming Social Science Teaching in Ukraine*, paper presented at the Second Annual CEP Eastern Scholars Roundtable, Lviv State University, Lviv, Ukraine, May 1999.
7. J. Elster, C. Offe, and U.K. Preuss, *Institutional Design in Post-Communist Societies: Rebuilding the Ship at Sea*, Cambridge: Cambridge University Press, 1998.
8. B. Readings, *University in Ruins*, Cambridge, Massachusetts, and London, England: Harvard University Press, 1996.
9. B.R. Clark, *Academic Power in Italy: Bureaucracy and Oligarchy in a National University System*, Chicago and London: The University of Chicago Press, 1977.
10. H. de Rudder, "The Transformation of East German Higher Education as Adaptation, Integration and Innovation," *Minerva*, 35 (1997), pp. 99–125.
11. V. Tomusk, *The Blinding Darkness of the Enlightenment: Towards the Understanding of Post State-Socialist Higher Education in Eastern Europe*, Turku: RUSE, 2000.

12. D. Levy, *Higher Education and the State in Latin America: Private Challenges to Public Dominance*, Chicago: The University of Chicago Press, 1986.
13. L. Balcerowicz, "Research and Education in the Post-Communist Transition," in *Western Paradigms and Eastern Agenda: A Reassessment*, Vienna: Institute of Human Sciences, 1995.
14. D. Tapper and D. Palfreyman, *Oxford and the Decline of Collegiate Tradition*, London and Portland, Oregon: Woburn Press, 2000.
15. V. Tomusk, "Enlightenment and Minority Cultures: Central and East European Higher Education Reform Ten Years Later," *Higher Education Policy*, 14, 1 (2001), pp. 61–73.
16. C. de Moura Castro and D. Levy, *Myth, Reality and Reform: Higher Education Policy in Latin America*, Washington, DC: The Johns Hopkins University Press, 2000.
17. Elster et al., *Institutional Design in Post-Communist Societies*.
18. Ibid.
19. J. Baudrillard, *The Illusion of the End*, Cambridge: Polity Press, 1994.
20. P. Drucker, "Defending Japanese Bureaucrats," *Foreign Affairs*, September/October 1998.
21. V. Tomusk, "The Syndrome of the Holy Degree: Critical Reflections on the Staff Development in Estonian Universities," *Higher Education Management*, 7, 3 (1995), pp. 385–397.
22. C. Offe, *Modernity and the State: East, West*, Cambridge, MA: The MIT Press, 1996.
23. Baudrillard, *The Illusion of the End*.
24. L.R. Graham, *Science, Philosophy, and Human Behavior in the Soviet Union*, New York: Columbia University Press, 1987.
25. Rudder, "The Transformation of East German Higher Education."
26. *Tertiary Education and Research in the Russian Federation*, Paris: OECD, 1999.
27. E. Gellner, "Enlightenment—Yes or No?" in Jozef Niznik and John T. Sanders (Eds.), *Debating the State of Philosophy: Habermas, Rorty, and Kolakowski*, Westport, Connecticut and London: Praeger, 1996.
28. T. Lajos, "Perspectives, Hopes and Disappointments: Higher Education Reform in Hungary," *European Journal of Education*, 28, 4 (1993), pp. 403–411.
29. R. Rorty, *Philosophy and Social Hope*, London: Penguin Books, 1999.
30. W.H. Newton-Smith, "Science and Open Society: Is the Scientific Community a Genuinely Open One?" in R. Dahrendorf, Y. Elkana, A. Neier, W. Newton-Smith, and I. Rév (Eds.), *The Paradoxes of Unintended Consequences*, Budapest and New York: Central European University Press, 2000.
31. N. Chomsky, "The Responsibility of Intellectuals," in James Peck (Ed.), *The Chomsky Reader*, New York: Pantheon Books, 1987.
32. P. Bourdieu, *Homo Academicus*, Cambridge: Polity Press, 1988, p. 114.
33. S. Žižek, *Did Somebody Say Totalitarianism? Five Interventions in the (Mis)use of a Notion*, London and New York: Verso, 2001.
34. O. Kivinen and S. Ahola, "Higher Education as Human Risk Capital: Reflections on Changing Labor Markets," *Higher Education*, 38, 2 (1999), pp. 191–208.
35. M. Kwiek, "The Nation-State, Globalization and the Modern Institution of the University," *Theoria: A Journal of Social and Political Theory*, 96 (2000), pp. 74–98.
36. Ibid.

Epilogue

1. E. Clapton, *My Father's Eyes*, http://www.eric-clapton.co.uk/ecla/lyrics/my-fathers-eyes.html.
2. I. Kant, *The Conflict of the Faculties*, Lincoln and London: University of Nebraska Press, 1992.

3. Ibid.
4. G. Neave, *Anything Goes: Or, How the Accommodation of Europe's Universities to European Integration Integrates an Inspiring Number of Contradictions*, keynote address delivered at the 23rd Annual Forum of EAIR—The European Higher Education Society, University of Porto, Porto, September 9–12, 2001.
5. Jeremiah 8:20.2, King James Authorized Version.
6. Neave, *Anything Goes*.

Bibliography

Aaviksoo, J. "Mõtteid ülikooli arengust (Thoughts on Development of the University)," *Tartu Ülikool*, 3, 4 (1992), pp. 39–44.
AI, *Report 1991–1995: The Hannah Arendt Prize 1996 Self-Study Questionnaire*, unpublished document, Bratislava: Academia Istropolitana, 1996.
AIN, *AINova—The Vision for Future*, unpublished document, Svjaty Jur: Academia Istropolitana Nova, 1999.
AIN, *Future Prospectus: Academia Istropolitana Nova*, unpublished document, Svjaty Jur: Academia Istropolitana Nova, 1998.
Al'ferov, Zh. I. and Sadovnichi, V.A. "Obrazovanie dlya Rossii XXI veka (Education for Russia in the 21st Century)." In: V.A. Sadovnichi (Red.), *Obrazovanie kotoroe my mozhem poteryat'* (*Education That We May Lose*), Moskva: Moskovskii gosudarstvennyi universitet im. M.V. Lomonosova, 2002.
Anderson, M. *Impostors in the Temple: A Blueprint for Improving Higher Education in America*, Stanford: Hoover Institution Press, 1996.
Apple, M.W. *Cultural Politics and Education*, Buckingham: Open University Press, 1996.
Balcerowicz, L. "Research and Education in the Post-Communist Transition," in *Western Paradigms and Eastern Agenda: A Reassessment*, Vienna: Institute of Human Sciences, 1995.
Bales, K. *Disposable People: New Slavery in the Global Economy*, Berkeley, Los Angeles, London: University of California Press, 1999.
Baron, S., Riddell, S., and Wilson, A. "The Secret of Eternal Youth: Identity, Risk and Learning Difficulties," *British Journal of Sociology of Education*, 20, 4 (1999), pp. 483–499.
Barrow, J.D. and Tipler, F.J. *The Anthropic Cosmological Principle*, Oxford and New York: Oxford University Press, 1986.
Baudrillard, J. *The Perfect Crime*, London and New York: Verso, 1996.
Baudrillard, J. *The Illusion of the End*, Cambridge: Polity Press, 1994.
Baudrillard, J. *Simulacres et Simulation*, Paris: Galilée, 1981.
Bauman, Z. "Pierre Bourdieu or the Dialectics of Vita Contemplativa and Vita Activa," *Revue International de Philosophie*, 56, 2 (2002), pp. 179–193.
Becher, T. *Academic Tribes and Territories: Intellectual Enquiry and the Cultures of Disciplines*, Buckingham and Bristol: Open University Press, 1989.
Berlin, I. "The Silence in Russian Culture," *Foreign Affairs*, October 1957.
Boia, L. *History and Myth in the Romanian Consciousness*, Budapest: Central European University Press, 2001.
Bollag, B. "Private Colleges Reshape Higher Education in Eastern Europe and Former Soviet States," *The Chronicle of Higher Education*, June 11, 1999.
Bologna Declaration, *The European Higher Education Area: Joint Declaration of the European Ministers of Education, Convened in Bologna on June 19, 1999*.
Bourdieu, P. *The State Nobility*, Stanford: Stanford University Press, 1996.
Bourdieu, P. *Homo Academicus*, Cambridge: Polity Press, 1988.
Bourdieu, P., Passeron, J.C., and Saint Martin, M. de, *Academic Discourse*, Cambridge: Polity Press, 1994

Brennan, J., Goedegebuure, L.C.J., Westerheijden, D.F., and Shah, T. *Comparing Quality in Europe*, publication no. 101 in the series Higher Education Policy Studies, Enschede: CHEPS, University of Twente, 1991.

Broadfoot, P. "Quality Standards and Control in Higher Education: What Price Life-Long Learning?" *International Journal of Sociology of Education*, 8, 2 (1998), pp.155–180.

Brooks, D. *Bobos in Paradise: The New Upper Class and How They Got There*, New York: Simon & Schuster, 2000.

Bruckbauer, S. "Ranking Productivity and Its Dynamics: Faltering Growth of Eastern Europe," *Economic Trends*, 3 (2000).

Buttgereit, M. "Higher Education and Its Relations to Employment in the USSR and in the Federal Republic of Germany: A Comparison." In: R. Avakov, M. Buttgereit, B.C. Sanyal, and U. Teichler (Eds.), *Higher Education and Employment in the USSR and in the Federal Republic of Germany*, Paris: IIEP, 1984, pp. 231–326.

Castoriadis, C. *The Imaginary Institution of Society*, Cambridge, Massachusetts: The MIT Press, 1987.

Cerych, L. "Educational Reforms in Central and Eastern Europe," *European Journal of Education*, 30, 4 (1995), pp. 423–435.

Cerych, L. and Sabatier, P. *Great Expectations and Mixed Performance: The Implementation of Higher Education Reforms in Europe*, Trentham: Trentham Books, 1986.

Chomsky, N. "The Responsibility of Intellectuals." In: James Peck (Ed.), *The Chomsky Reader*, New York: Pantheon Books, 1987.

Chomsky, N. *The Old and the New Cold War*, New York: Pantheon Books, 1980.

Chomsky, N. *Language and Freedom*, New York: Pantheon Books, 1970.

Chuprunov, D., Avakov, R., and Jiltsov, E. "Higher Education, Employment and Technological Progress in the USSR." In: R. Avakov, M. Buttgereit, B.C. Sanyal, and U. Teichler (Eds.), *Higher Education and Employment in the USSR and in the Federal Republic of Germany*, Paris: UNESCO/IIEP, 1984.

Clapton, E. *My Father's Eyes*, http://www.eric-clapton.co.uk/ecla/lyrics/my-fathers-eyes.html. Site visited on October 11, 2003.

Clark, B.R. *Creating Entrepreneurial Universities: Organizational Pathways of Transformation*, London: Pergamon Press, 1998.

Clark, B.R. *Places of Inquiry: Research and Advanced Education in Modern Universities*, Berkeley and Los Angeles, California: University of California Press, 1995.

Clark, B.R. *The Higher Education System: Academic Organization in Cross-National Perspective*, Berkeley and Los Angeles: University of California Press, 1983.

Clark, B.R. *Academic Power in Italy: Bureaucracy and Oligarchy in a National University System*, Chicago and London: The University of Chicago Press, 1977.

CoE, *Report of the Advisory Mission to Slovenia: The Draft Law on Organisation and Financing in the Field of Science and Technology, 21–24 February 1996*, Council of Europe Document DECS LRP 96/07. Strasbourg: Council of Europe.

CoE, *Croatia: Report of the Advisory Mission on Quality Assurance, Zagreb, 15-17 March, 1995*, Council of Europe Document DECS LRP 95/07. Strasbourg: Council of Europe, 1995.

CoE, *Regulation of Private Higher Education: Report of the Multilateral Workshop, Prague, 9–11 May 1994*, Council of Europe Document DECS LRP (94) 33. Strasbourg: Council of Europe.

CoE, *Estonia: Report of The Advisory Mission, 18–21 May 1994*, Council of Europe Document DECS LRP 94/11. Strasbourg: Council of Europe.

CoE, *Universities, Colleges and Others: Diversity of Structures for Higher Education: Report of the Multilateral Workshop, Bucharest, 23–25 September 1993*, Council of Europe Document DECS LRP (94) 05. Strasbourg: Council of Europe.

CoE, *Brain Drain from Universities: Final Report of the CC-PU Forum Role Conference*, Council of Europe Document DECS-HE 94/26. Strasbourg: Council of Europe.

Colbeck, C. "Reshaping Forces that Perpetuate the Research/Practice Gap: Focus on New Faculty." In: A. Kezar and P. Eckel (Eds.), *Higher Education Research: Transfer to Practice. New Directions for Higher Education*, San Francisco: Jossey-Bass, 2000.

Coleman, J.S. "Social Capital in the Creation of Human Capital." In: P. Dasgupta and I. Serageldin (Eds.), *Social Capital: A Multifaceted Perspective*, Washington, DC: The World Bank, 2000.

Collins, R. *The Sociology of Philosophies*, Cambridge, Massachusetts, and London, England: The Belknap Press of Harvard University Press, 1998.

Coopers and Lybrand. *Hungarian Higher Education: Content and Development of HEI Business Plans. Final Report for the World Bank*, unpublished document, 1997.

Coyne, R. *Technoromanticism: Digital Narrative, Holism, and the Romance of the Real*, Cambridge, Massachusetts and London, England: The MIT Press, 1999.

Dahrendorf, R. *Universities After Communism: The Hannah Arendt Prize and the Reform of Higher Education in East Central Europe*, Hamburg: Edition Körber-Stiftung, 2000.

Davis, E. *Techgnosis: Myth, Magic and Mysticism in the Age of Information*, London: Serpent's Tail, 1999.

Darvas, P. "Institutional Innovation in Central and East European Higher Education." In: M. Szymonski, and I. Guzik (Eds.), *Research at Central and East European Universities*, Krakow: Jagellonian University Press, 1997, pp. 129–149.

Darvas, P. *Institutional Innovation in Central European Higher Education*, Vienna, Institute for Human Sciences, 1996.

David-Fox, M. *Revolution of the Mind: Higher Learning Among the Bolsheviks, 1918–1929*, Ithaca and London: Cornell University Press, 1997.

Davies, J.L. *The Entrepreneurial and Adaptive University*, Paris: OECD, 1987.

Drucker, P. "Defending Japanese Bureaucrats," *Foreign Affairs*, September/October 1998.

Egorshin, A.P. "Perspektivy razvitiia obrazovaniia Rossii v XXI v. (Perspectives on the Development of Education in Russia in the 21st Century)," *Universitetskoe Upravlenie: Praktika i Analiz (University Management: Practice and Analysis)*, 4 (2000), pp. 50–64.

EHU, *The Current State and Prospects for the Development of the European Humanities University*, unpublished report, Minsk: EHU, 1998.

EHU, *Uchreditel'nyi dogovor o sozdanii i deyatel'nosti evropeiskogo gumanitarnogo universiteta (Founding Agreement on the Establishment and Activities of the European Humanities University)*, unpublished document, Minsk: EHU, 1997.

EHU, *Ustav Evropeiskogo Gumanitarnogo Universiteta (Charter of the European Humanities University)*, unpublished document, Minsk: EHU, 1997.

Einhorn, J. "The World Bank's Mission Creep," *Foreign Affairs*, September/October 2001, pp. 22–35.

Elster, J., Offe, C., and Preuss, U.K. *Institutional Design in Post-communist Societies: Rebuilding the Ship at Sea*, Cambridge: Cambridge University Press, 1998.

ER, Kõrgharidusreform aastatel 2001–2002 (Higher Education Reform in the Years 2001–2002), *Eesti Vabariigi Valitsuse käskkiri, June 12, 2001 (Decree of Government of the Republic of Estonia*, June 12, 2001).

ER, Ülikooliseadus (The University Law), *Riigi Teataja* (1995), pp. 12, 119; revisions: (1996), pp. 49, 953; (1996), pp. 51, 965; (1997), pp. 42, 678; (1999), pp. 10, 150.

ER, Rakenduskõrgkooli seadus (The Law of Vocational Higher Education Institutions), *Riigi Teataja* (1998), pp. 61, 980.

ER, Erakooliseadus (The Law of the Private School), *Riigi Teataja* (1998), pp. 57, 859.

ER, Teadus—ja arendustegevuse korralduse seadus (The Law of the Organizing the Research and Development Activities), *Riigi Teataja* (1997), pp. 30, 471.

ER, Erakooliseadus (The Law of the Private School), *Riigi Teataja* (1993), pp. 35, 547, revisions: (1995), pp. 12, 119; (1996), pp. 49, 953, 51, 965.

ER, Kõrghariduse hindamise nõukogu põhikiri (Statute of the Higher Education Evaluation Council), *Kultuuri-ja Haridusministeerium Teataja (Information Bulletin of the Ministry of Culture and Education)*, 1995.
ER, Tartu Ülikooli seadus (The Law of the University of Tartu), *Riigi Teataja* (1995), pp. 23, 333.
ER, Tsiviilseadustiku üldosa seadus (The Law of the General Part of the Civil Code), *Riigi Teataja*, 1994, pp. 53, 889.
ER, Erakooliseadus (The Law of Private Educational Institutions), *Riigi Teataja* (State Information Bulletin), 1993, pp. 35, 547.
ER, *Eesti Vabariigi rakendusliku kõrgkooli ajutine põhimäärus, Lisa 1 EV Haridusministeeriumi määrusele (Temporary Edict of the Vocational Higher Education Institutions of the Republic of Estonia, Appendix 1 to the Decree of the Ministry of Education of the RE)*, No. 4, December 19, 1991 (unpublished document).
ESSR, *Eesti NSV Haridusministeeriumi käskkiri (Decree of the Ministry of Education of the ESSR)*, No. 156, June 6, 1990.
ESSR, *ENSV Rükliku Hariduskomitee ja ENSV Kehakultuuri ja Spordikomitee käskkiri (Decree of the State Education Committee of the ESSR and the Committee for Sports of the ESSR)*, No. 286–r287, June 29, 1989.
ESSR, *Lisa 2. ENSV Riikliku Hariduskomitee kolleegiumi otsus (Appendix 2. Decision of the Council of the State Education Committee of the ESSR)*, No. 8–5, June 22, 1989.
Feyerabend, P. *Three Dialogues on Knowledge*, Oxford, UK and Cambridge, USA: Blackwell, 1991.
Frederiks, M.M.H., Westerheijden, D.F., and Weusthof, P.J.M. "Effects of Quality Assessment in Dutch Higher Education," *European Journal of Education*, 29, 2 (1994), pp. 181–199.
Frydman, R., Murphy, K., and Rapaczynski, A. *Capitalism with a Comrade's Face*, Budapest: Central European University Press, 1998.
Gaddy, C. and Ickes, B. "Russia's Virtual Economy," *Foreign Affairs*, September/October 1998.
Gale, P. "Indigenous Rights and Higher Education in Australia: 'Not Just Black and White.'" In: B. Teasdale and Z. Ma Rhea (Eds.), *Local Knowledge and Wisdom in Higher Education*, Oxford: Pergamon, 2000.
Galko, T.E. *Belorusskii kommercheskii universitet upravlenia (The Belarus Commercial University of Management)*, unpublished report, Minsk: BCUM, 1997.
Geiger, R.L. *Single Donor Universities*, PONP Working Paper No. 215 and ISPS Working Paper No. 2215, Yale University, Institution for Social and Policy Studies, 1995.
Gellner, E. *Nationalism*, London: Phoenix, 1998.
Gellner, E. "Enlightenment—Yes or No?" In: J. Niznik and J.T. Sanders (Eds.), *Debating the State of Philosophy: Habermas, Rorty, and Kolakowski*, Westport, Connecticut and London: Praeger, 1996.
Gellner, E. *Conditions of Liberty: Civil Society and Its Rivals*, New York: Allen Lane/Penguin, 1994.
Getty, J.A. and Naumov, O.V. *The Road to Terror: Stalin and the Self-Destruction of the Bolsheviks, 1932–1939*, New Haven and London: Yale University Press, 1999.
Gibbons, M., Limoges, C., Nowotny, H., Schwartzman, S., Scott, P., and Trow, M. *The New Production of Knowledge: The Dynamics of Science and Research in Contemporary Societies*, London, Thousand Oaks, New Delhi: Sage Publications, 1994.
Giddens, A. *The Third Way: The Renewal of Social Democracy*, Cambridge: Polity Press, 1998.
Gilder, E. *Report on the UNESCO/CEPES Seminar "Quality Management in Higher Education," Bucharest, May 23, 1996*, unpublished document.
Godet, M. *From Anticipation to Action: A Handbook of Strategic Prospective*, Paris: UNESCO, 1993.
Goduka, I.N. "African/Indigenous Philosophies: Legitimizing Spiritually Centred Wisdoms Within the Academy." In: P. Higgs, N.C.G. Vakalisa, T.V. Mda, and N.T. Assié-Lumumba (Eds.), *African Voices in Education*, Lansdowne: Juta, 2000.

Gorbachev, M.S. *Perestroika i novoe myshlenie dlia nashei strany i dlia vsego mira* (*Perestroika and New Thinking for Our Country and for the Whole World*), Moscow: Politizdat, 1987.
Graham, L.R. *Science, Philosophy, and Human Behavior in the Soviet Union*, New York: Columbia University Press, 1987.
Habermas, J. *The New Conservatism*, Cambridge: Polity Press, 1989.
Hancock, G. *Lords of Poverty: The Power, Prestige, and Corruption of the International Aid Business*, New York: Atlantic Monthly Press, 1989.
Hansen, C.R. In: *United States-Soviet Relation: 1988, Hearings Before the Subcommittee on the Europe and Middle East of the Committee on Foreign Affairs, House of Representatives, One Hundredth Congress*, Vol. I, Washington: US Government Printing Office, 1988, p. 432.
Hatar, V. *The Right to Sanity*, Budapest: Corvina Books Ltd., 1999.
Havel, I. *Living in Conceivable Worlds*, paper presented at the First World Congress of Paraconsistency, Ghent, July 30–August 2, 1997.
Hayek, F.A. *The Intellectuals and Socialism*, London: IEA Health and Welfare Unit, 1949/98.
Heller, A. *The Three Logics of Modernity and the Double Bind of the Modern Imagination*, public lecture series no. 23, Budapest: Collegium Budapest, 2000.
Henry, M., Lingard, B., Rizvi, F., and Taylor, S. *The OECD, Globalization and Education Policy*, Oxford: Pergamon, 2001.
Herrnstein, R.J. and Murray, C. *The Bell Curve: Intelligence and Class Structure in American Life*, New York, London, Toronto, Tokyo, Singapore: Free Press, 1994.
Heyneman, S.P. "The Transition from Parry State to Open Democracy: The Role of Education," *International Journal of Educational Development*, 2000.
Hickling Hudson, A. "The Cuban University and Educational Outreach: Cuba's Contribution to Post-Colonial Development." In: B. Teasdale and Z. Ma Rhea (Eds.), *Local Knowledge and Wisdom in Higher Education*, Oxford: Pergamon, 2000.
Hoogvelt, A. *Globalization and the Postcolonial World: The New Political Economy of Development*, Second edition, Hampshire: Palgrave, 2001.
Hountondji, P.J. "Manufacturing Unemployment: The Crisis of Education in Africa." In: P. Higgs, N.C.G. Vakalisa, T.V. Mda, and N.T. Assié-Lumumba (Eds.), *African Voices in Education*, Lansdowne: Juta, 2000.
Iliescu, I. Opening speech given at the international symposium, "*Central Europe—South-Eastern Europe: Inter-regional Relational Relations in the Field of Education, Science, Culture, and Communication,*" Bucharest, Romania, April 19–22, 2001.
IHF, *International Humanitarian Foundation*, Minsk: IHF, 1998.
in't Veld, R., Füssel, H.-P., and Neave, G. (Eds.), *Relations Between State and Higher Education*, The Hague: Kluwer Law International, 1996.
Johnes, G. "The Funding of Higher Education in the United Kingdom." In: P. Hare (Ed.), *Structure and Financing of Higher Education in Russia, Ukraine and the EU*, London: Jessica Kingsley Publishers, 1997.
Jones, E.A. and Ratcliff, J.L. "Global Perspectives on Program Assessment and Accreditation." In: L. Lategan, M. Fourie, and A. Strydom (Eds.), *Programme Assessment in Higher Education in South Africa*, Blomfontein, RSA: Unit for Higher Education Research, University of the Orange Free State, 1999.
Kaf, *1997 Yearbook of the Kyrgyz American Faculty*, Bishkek: Kyrgyz-American School, Kyrgyz State National University, 1997.
Kant, I. *The Conflict of the Faculties*, Lincoln and London: University of Nebraska Press, 1992.
Kazmin, A. "Rush to Plug Vietnam's Learning Gap," *The Financial Times*, July 2, 2002.
Kent, R. *Two Positions in the International Debate About Higher Education: The World Bank and UNESCO*, paper presented at the 1995 meeting of the Latin American Studies Association. Washington, DC, September 28–30, 1995.
Kinelyev, V. *Preface to State Educational Standard of Higher Professional Education*, Moscow: Goskomvuz, 1995.

Kivinen, O. and Ahola, S. *Higher Education as Human Risk Capital: Reflections on Changing Labor Markets*, Higher Education, 38, 2 (1999), pp. 191–208.

Kivinen, O. and Rinne, R. "State, Governmentality and Education—the Nordic Experience," *British Journal of Sociology of Education*, 19 (1998), pp. 39–52.

Korka, M. *Strategy and Action in the Reform of Education in Romania*, Bucharest: Paideia, 2000.

Kovaleva, E. *Progress and Issues in Reforming Social Science Teaching in Ukraine*, paper presented at the Second Annual CEP Eastern Scholars Roundtable, Lviv State University, Lviv, Ukraine, May 1999.

Kovaleva, N. "Women and Engineering Training in Russia," *European Journal of Education*, 34, 4 (1999), pp. 425–435.

Kroos, K. *Why There is so Little Reform in Estonian Higher Education: Dual Identity of Estonian Intellectuals*, unpublished M.A. thesis, Budapest, Central European University, 2000.

Kuhn, T. *The Structure of Scientific Revolutions*, second edition, enlarged, Chicago and London: University of Chicago Press, 1962.

Kwiek, M. "The Nation-State, Globalization and the Modern Institution of the University," *Theoria: A Journal of Social and Political Theory*, 96 (2000), pp. 74–98.

Lajos, T. "Perspectives, Hopes and Disappointments: Higher Education Reform in Hungary," *European Journal of Education*, 28, 4 (1993), pp. 403–411.

Lane, D. *The Rise and Fall of State Socialism*, Oxford: Polity Press, 1996.

Lao-tzu, *Te-Tao Ching*, New York: The Modern Library, 1993.

Leary, T. *Interpersonal Diagnosis of Personality*, New York: John Wiley & Sons, 1957.

Lem, S. *The Star Diaries*, San Diego, New York, London: A Harvest Book, 1985.

Lenin, V.I. "Letter to Maxim Gorky, 15 September, 1919." In: D. Koenker and R. Bachman (Eds.), *Revelations from the Russian Archives*, Washington, DC: Library of Congress, 1997.

Lennon, J. *The John Lennon Collection*, London: EMI Records, 1989.

Levy, D. *Public Policy for Hungarian Private Higher Education*, unpublished report, 1997.

Levy, D. *Higher Education and the State in Latin America: Private Challenges to Public Dominance*, Chicago: The University of Chicago Press, 1986.

Lilly, J.C. *The Scientist: A Metaphysical Autobiography*, Berkeley, CA: Ronin Publishers, 1996.

Lukacs, J. *At the End of an Age*, New Haven & London: Yale University Press, 2002.

Lumumba-Kasongo, T. "Rethinking Educational Paradigms in Africa: Imperatives for Social Progress in the New Millennium." In: P. Higgs, N.C.G. Vakalisa, T.V. Mda, and N.T. Assié-Lumumba (Eds.), *African Voices in Education*, Lansdowne: Juta, 2000.

Maior, L. and Georgescu, L. *Status of the Higher Education Reform in Romania*, unpublished report for the World Bank, 1996.

Marcuse, H. *One-Dimensional Man: Studies in the Ideology of Advanced Industrial Society*, London: Routledge, 1964/91.

Marx, K. and Engels, F. "Manifesto of the Communist Party." In: Karl Marx and Frederick Engels, *Economic and Philosophic Manuscripts of 1844 and the Communist Manifesto*, New York: Prometheus Books.

Matos, N. "The Nature of Learning, Teaching and Research in Higher Education in Africa." In: P. Higgs, N.C.G. Vakalisa, T.V. Mda, and N.T. Assié-Lumumba (Eds.), *African Voices in Education*, Lansdowne: Juta, 2000.

Matthews, M. *Patterns of Deprivation in the Soviet Union Under Brezhnev and Gorbachev*, Stanford: Hoover Institution Press, 1989.

Matthews, M. *Education in the Soviet Union: Politics and Institutions Since Stalin*, London: George Allen & Unwin, 1982.

McLuhan, M. *War and Peace in the Global Village*, San Francisco: HardWired, 1968/97.

McLuhan, M. *Understanding Media: The Extension of Man*, Cambridge, Massachusetts and London, England: The MIT Press, 1964/94.

Merton, R.K. *Social Theory and Social Structure*. New York: Free Press, 1968.
Mokhov, V.P. "Stratifikaciya sovetskoi regional'noj politicheskoj elity. 1960–1990 gg. (Stratification of the Soviet Regional Political Elite, 1960s–1990s)." In: M.N. Afanas'ev (Ed.), *Vlast' i obshchestvo v postsovetskoi Rossii: novye praktiki i instituty* (*Power and Society in Post-Soviet Russia: New Practices and Institutions*), Moscow: Moscow Public Science Foundation, 1999.
More, M. *The Extropian Principles: A Transhumanist Declaration*. Version 3.0, available in the Internet: http://www.extropy.com/extprn3.htm.
Morgan, A.W. and Bergerson, A.A. "Importing Organizational Reform: The Case of Lay Boards in Hungary," *Higher Education*, 40 (2000), pp. 423–448.
Motyl, A.J. "After Empire: Competing Discourses an Inter-State Conflict in Post Imperial Eastern Europe." In: B.R. Rubin and J. Snyder (Eds.), *Post-Soviet Political Order*, London: Routledge, 1998.
Moura Castro, C. de and Levy, D.C. *Myth, Reality and Reform: Higher Education Policy in Latin America*, Washington, DC: Inter-American Development Bank, 2000.
Moynihan, D.P. "The Professionalization of Reform," *The Public Interest*, 1 (1965), pp. 6–16.
Murakami, H. *The Wind-up Bird Chronicle*, London: The Harvill Press, 1998.
Murphy, J.A. "Degree of Waste: The Economic Benefits of Educational Expansion," *Oxford Review of Education*, 19, 1 (1993), pp. 9–31.
MVShSEN. *Obrazovatel'naya politika i obrazovatel'noe sakonodatel'stvo v sovremennoi Rossii. Statisticheskie dannye* (*Educational Policy and Educational Legislation in Contemporary Russia. Statistical Data*), Moskva: MVShSEN, 2002.
Naude, J.P. "La science russe survit tant bien que mal d'expédients," *Le Monde*, October 16, 1996.
Neave, G. *Anything Goes: Or, How the Accommodation of Europe's Universities to European Integration Integrates an Inspiring Number of Contradictions*, keynote address delivered at the 23rd Annual Forum of EAIR—The European Higher Education Society, University of Porto, Porto, September 9–12, 2001.
Neave, G. "On Living in Interesting Times: Higher Education in Western Europe 1985–1995," *European Journal of Education*, 30 (1995), pp. 377–393.
Neave, G. and VanVught, F. *Prometheus Bound: The Changing Relationship Between Government and Higher Education in Western Europe*, Oxford, New York, Beijing, Frankfurt, São Paulo, Sydney, Tokyo, Toronto: Pergamon Press, 1991.
Neumann, I.B. *Uses of the Other: "The East" in European Identity Formation*, Manchester: Manchester University Press, 1999.
Newsweek, "Moldova: Scarred For Life," *Newsweek*, July 16, 2001.
Newton-Smith, W.H. "The Origin of the Universe," In: P.J.N. Baert (Ed.), *Time in Contemporary Intellectual Thought*, Amsterdam, Lausanne, New York, Oxford, Shannon, Singapore, Tokyo: Elsevier Science B.V., 2000.
Newton-Smith, W.H. "Science and Open Society: Is the Scientific Community a Genuinely Open One?" In: R. Dahrendorf, Y. Elkana, A. Neier, W. Newton-Smith, and I. Rév (Eds.), *The Paradoxes of Unintended Consequences*, Budapest and New York: Central European University Press, 2000.
Nicolescu, L. *Private vs. State Higher Education in Romania: The Business Community Perspective*, unpublished manuscript.
Nkomo, M. "Educational Research in the African Development Context: Rediscovery, Reconstruction and Prospects." In: P. Higgs, N.C.G. Vakalisa, T.V. Mda, and N.T. Assié-Lumumba (Eds.), *African Voices in Education*, Lansdowne: Juta, 2000.
Odora Hoppers, C.A. "African Voices in Education: Retrieving the Past, Engaging the Present and Shaping the Future." In: P. Higgs, N.C.G. Vakalisa, T.V. Mda, and N.T. Assié-Lumumba (Eds.), *African Voices in Education*, Lansdowne: Juta, 2000.
OECD, *Reviews of National Policies for Education: Romania*, Paris: OECD Publications, 2000.

OECD, *Education and Learning for Sustainable Consumption*, OECD document OM/ENV/CERI (99) 64, Paris: OECD, 1999.
OECD, *Tertiary Education and Research in the Russian Federation*, Paris: OECD, 1999.
OECD, *Review of Higher Education in the Czech and Slovak Federal Republic: Examiner's Report and Questions*, OECD document DEELSA/ED/WD (92) 5, Paris: OECD, 1992.
O'Hear, A. *After Progress: Finding the Old Way Forward*, London: Bloomsbury, 1999.
Offe, C. "Democracy and Trust," *Theoria: A Journal of Social and Political Theory*, 96 (2000), pp. 1–13.
Offe, C. *Modernity and the State: East, West*, Cambridge, Massachusetts, and London, England: The MIT Press, 1996.
Pavlychko, S., et al. *European Humanities University: Report of the Review Panel*, unpublished report of HESP/OSI, 1998.
Petrova, L.E. *Novye bednye uchenye: zhisnennye strategii v usloviyakh krizisa* (*New Poor Scholars: Life Strategies Under the Conditions of a Crisis*), paper presented at the Conference Russian Social Sciences: A New Perspective, Moscow, October 7–9, 1999.
Pirsig, R.M. *Zen and the Art of Motorcycle Maintenance: an Inquiry into Values*, London: Vintage, 1974.
Piskunov, D. "Russia: Higher Education and Change." In: A.D. Tillett and B. Lesser (Eds.), *An Uncertain Transition: Preliminary Assessment of Higher Education, Science and Technology in Central and Eastern Europe*, Halifax: Dalhousie University, Lester Pearson Institute, 1993.
Plato. "Socrates Defence (Apology)." In: E. Hamilton and H. Cairns (Eds.), *Plato: The Collected Dialogues*, Princeton: Princeton University Press, 1961.
Polanyi, M. *Science, Faith and Society*, Chicago and London: University of Chicago Press, 1946.
Popper, K.R. *Alles Leben ist Problemlösen: Über Erkenntnis, Geschichte und Politik*, München, Zürich: Piper, 1994.
Poster, M. "Cyberdemocracy: The Internet and the Public Sphere." In: D. Trend (Ed.), *Reading Digital Culture*, Malden, MA: Blackwell Publishers, 2001.
Prifti, I. "Albania." In: R. in't Veld, H.-P. Füssel, and G. Neave (Eds.), *Relations Between State and Higher Education*, The Hague: Kluwer Law International, 1996.
Putnam, H. *Words and Life*, Cambridge, Massachusetts, and London, England: Harvard University Press, 1994.
Ramonet, I. "L'axe du Mal," *Le Monde Diplomatique*, March 2002.
Ratcliff, J.L. "Institutional Self-Evaluation and Quality Assurance: A Global View." In: A. Strydom and L. Lategan (Eds.), *Institutional Self-Evaluation in Higher Education in South Africa*, Blomfontein, RSA: Unit for Higher Education Research, University of the Orange Free State, 1998.
Readings, B. *University in Ruins*, Cambridge, Massachusetts, and London, England: Harvard University Press, 1996.
Rév, I. "The-Self-Not-Fulfilling Prophecy." In: R. Dahrendorf, Y. Elkana, A. Neier, W. Newton-Smith, and I. Rév (Eds.), *The Paradoxes of Unintended Consequences*, Budapest and New York: Central European University Press, 2000.
RF, Grazhdanskii Kodeks Rossiiskoi Federacii. Oficial'nyi Tekst (The Civil Code of the Russian Federation. Official Text), Moskva: Ekzamen, 2001.
RF, *Natsional'naya doktrina obrazovaniya v Rossiiskoi Federatsii: proekt* (*National Doctrine of Education of the Russian Federation: Draft*), document drafted by the Science and Education Committee of the State Duma of the Russian Federation, 1999.
RF, Zakon ob obrazowanii, Rossiiskoi Federacii (The Law on Education of the Russian Federation), *Rossijskaya Cazeta*, January 26, 1996.
RF, *Podgotovka specialistov v oblasti gumanitarnykh i social'no-ekonomicheskikh nauk* (*Training of Experts in Humanities and Economics*), Moskva: Goskomvuz, 1995.

RF, *Gosudarstvennyi Obrazovatelnyi Standart Vysshego Professionalnogo Obrazovaniia: Gosudarstvenniie Trebovaniia k minimumu soderzhaniia i urovniu podgotovki vypustnika po specialnosti 071900: Informatsionniie Sistemy v Ekonomike* (*State Educational Standard for a Minimum Content and Level of Training of Graduates from the Programme 071900: Informational Systems in Economics*), Moscow: Goskomvuz, 1995.
Romania, *Government Program*, Bucharest: Government of Romania, 2000.
Romania, *Declaration*, Bucharest: Government of Romania, 2000.
Romania, *The National Medium-Term Development Strategy of the Romanian Economy*, Bucharest: Government of Romania, 2000.
Romania, *Higher Education in a Learning Society: Argument for a New National Policy on the Sustainable Development of Higher Education*, Bucharest: The Ministry of National Education, 1998.
Romania, *Carte Blanche of the Reform of Education in Romania*, Bucharest: Ministry of Education, 1995.
Romania, *The Law on the Accreditation of Higher Education Institutions and the Recognition of Diplomas*, Bucharest: The Parliament of Romania, 1993.
Roper, S.D. *Romania: The Unfinished Revolution*, Amsterdam: Harwood Academic Publishers, 2000.
Rorty, R. *Philosophy and Social Hope*, London: Penguin Books, 1999.
Rudder, H., de. "The Transformation of East German Higher Education as Adaptation, Integration and Innovation," *Minerva*, 35 (1997), pp. 99–125.
Sadlak, J. "The Emergence of a Diversified System: The State/Private Predicament in Transforming Higher Education in Romania," *European Journal of Education*, 29, 1 (1994), pp. 13–23.
Sadovnichi, V.A. (Red.). *Obrazovanie kotoroe my mozhem poteryat'* (*Education That We may Lose*), Moskva: Moskovskii gosudarstvennyi universitet im. M.V. Lomonosova, 2002.
Salmi, J. "Globalisation and the Knowledge-Based Economy Pose Challenges to all Purveyors of Tertiary Education. Tertiary Education—News Item," *Times Higher Education Supplement*, December 3, 1999.
Sapatoru, D. *Public or Private? Post-Secondary Education Choices in Romania*, unpublished manuscript.
Savchuk, V., Luzik, P., Gal, I., and Oparin, V. "Higher Education in Ukraine: Structure and Financing," In: P. Hare (Ed.), *Structure and Financing of Higher Education in Russia, Ukraine and the EU*, London: Jessica Kingsley Publishers, 1997.
Schutz, A. *On Phenomenology and Social Relations*, Chicago: University of Chicago Press, 1975.
Seepe, S. "Africanization of Knowledge: Exploring Mathematical and Scientific Knowledge Embedded in African Cultural Practices." In: P. Higgs, N.C.G. Vakalisa, T.V. Mda, and N.T. Assié-Lumumba (Eds.), *African Voices in Education*, Lansdowne: Juta, 2000.
Sekirinskii, S.S. (Ed.). "Sovietskoe Proshloe: poiski ponimaniia (Soviet Past: Search for Understanding)," *Otechestvennaia Istoriia* (*Fatherland's History*), 4 (2000).
Shenk, D. *Data Smog: Surviving the Information Glut*, New York: HarperEdge, 1997.
Skillbeck, M. *The Challenge of Diversifying Tertiary Education: OECD Responses*, paper presented at the World Bank/Open Society Institute Workshop on Diversification of Higher Education in Central and Eastern Europe, Budapest, November 18–20, 1999.
Sklair, L. *The Transnational Capitalist Class*, Oxford and Malden: Blackwell Publishers, 2001.
Slantcheva, S. *The Challenges to Vertical Degree Differentiation Within Bulgarian Universities: The Problematic Introduction of the Three-Level System of Higher Education*, paper presented at the 21st Annual EAIR Forum, Lund, Sweden, August 26–29, 1999.
Sokolov, A.K. and Tyazhel'nikova, V.S. *Kurs Sovetskoi Istorii 1941–1991* (*The Course of the Soviet History 1941–1991*), Moskva: Vysshaia Shkola, 1999.

Solow, R.M. "Notes on Social Capital and Economic Performance." In: P. Dasgupta and I. Serageldin (Eds.), *Social Capital: A Multifaceted Perspective*, Washington, DC: The World Bank, 2000.

Soros, G. *Open Society: Reforming Global Capitalism*, London: Little, Brown and Company, 2000.

Stepko, M. "Ukraine." In: R. In 't Veld, H.-P. Füssel, and G. Neave (Eds.), *Relations Between State and Higher Education*, The Hague: Kluwer Law International, 1996.

Sterian, P. *Quality Assurance System in Romanian Higher Education*, paper presented at the Regional Training Seminar for Quality Assurance in Higher Education: Self Assessment and Peer Review, Budapest, November 10–16, 1996.

St. Petersburg State Technical University, *Koncepcii, Struktury i Soderzhanie Mnogourovnei Sistemy Vysshego Technicheskogo Obrazovaniia Rossii* (*Concepts, Structures and Content of Multilevel System of Higher Technical Education at Russia*), St Petersburg: State Technical University, 1993.

Sztompka, P. "Conceptual Frameworks in Comparative Inquiry: Divergent or Convergent?" In: M. Albrow and E. King (Eds.), *Globalization, Knowledge and Society*, London, Thousand Oaks, New Delhi: Sage Publications, 1990.

Tannehill, M. and Tannehill, L. *The Market for Liberty*, New York: Libertarian Review Foundation, 1984.

Tapper, D. and Palfreyman, D. *Oxford and the Decline of Collegiate Tradition*, London and Portland, Oregon: Woburn Press, 2000.

Tayler, J. "Russia is Finished. The Unstoppable Descent of a Once Great Power into Social Catastrophe and Strategic Irrelevance," *Atlantic Monthly*, May 2001.

Teffo, L.J. "Africanist Thinking: An Invitation to Authenticity." In: P. Higgs, N.C.G. Vakalisa, T.V. Mda, and N.T. Assié-Lumumba (Eds.), *African Voices in Education*, Lansdowne: Juta, 2000.

Teichler, U. *Changing Patterns of the Higher Education System: Experience of Three Decades*, London: Jessica Kingsley Publishers, 1988.

Todd, E. *The Final Fall: An Essay on the Decomposition of the Soviet Sphere*, New York: Karz Publishers, 1979.

Tomusk, V. *Ministries of Education and Higher Education Policies in Eastern Europe—Steering from Where?* UNESCO document IIEP/SEM 199, Paris: UNESCO, 2001, presented at the Policy Forum of the Organization of Ministries of Education, IIEP, Paris, June 20–21, 2001.

Tomusk, V. "Enlightenment and Minority Cultures: Central and East European Higher Education Reform Ten Years Later," *Higher Education Policy*, 14, 1 (2001), pp. 61–73.

Tomusk, V. *The Blinding Darkness of the Enlightenment: Towards the Understanding of Post State-Socialist Higher Education in Eastern Europe*, Turku: RUSE, 2000.

Tomusk, V. "When West Meets East: Decontextualizing the Quality of East European Higher Education," *Quality in Higher Education*, 6, 3 (2000).

Tomusk, V. "Reproduction of the 'State Nobility' in Eastern Europe: Past Patterns and New Practices," *British Journal of Sociology of Education*, 21, 2 (2000), pp. 269–283.

Tomusk, V. "Developments in Russian Higher Education: Legislative and Policy Reform Within Central and East European Context," *Minerva*, 36, 2 (1998), pp. 125–146.

Tomusk, V. "Estonia: Higher Education System." In: G. Neave, et al. (Eds.), *Complete Encyclopaedia of Education/Encyclopaedia of Higher Education on CD-ROM*, London: Elsevier Science, 1998.

Tomusk, V. "External Quality Assurance in Estonian Higher Education: Its Glory, Take-Off and Crash," *Quality in Higher Education*, 3 (1997), pp. 173–181.

Tomusk, V. "Between Politics and Professionalism: Reforming Fundamentals of Higher Education in Central and East Europe." In: T. Thanasuthipitak and S.L. Rieb (Eds.), *A Blueprint for Better Graduate Studies*, Chiang Mai: Graduate School, Chiang Mai University, 1997, pp. 331–343.

Tomusk, V. "Conflict and Interaction in Central and East European Higher Education: The Triangle of Red Giants, White Dwarfs and Black Holes," *Tertiary Education and Management*, 3, 3 (1997), pp. 247–255.

Tomusk, V. "Recent Trends in Estonian Higher Education: Emergence of the Binary Division from the Point of View of Staff Development," *Minerva*, 34, 3 (1996), pp. 279–289, 1996.

Tomusk, V. "Quality in Transition: Attributing a Meaning to Quality in Central-East European Higher Education." In: J.L. Lambert and T.W. Banta (Compilers), *Proceeding of the Eighth International Conference on Assessing Quality in Higher Education, July 14–16, 1996, Queensland, Australia*, Indiana: Indiana University-Purdue University Indianapolis, 1996.

Tomusk, V. "Discovering the Terra Incognita: The Changing Legal Landscape of Higher Education in Estonia." In: *The World on the Move and Higher Education in Transition: Selected Papers from the Conference Organised by the Programme on Institutional Management in Higher Education (IMHE) of the OECD. Central European University, Centre for Higher Education Studies in Prague, Prague, Czech Republic, August 23–25, 1995*, pp. 147–154.

Tomusk, V. "Nobody can Better Destroy Your Higher Education than Yourself: Critical Remarks About Quality Assessment and Funding in Estonian Higher Education," *Assessment and Evaluation in Higher Education*, XX, 1 (1995), pp. 115–124.

Tomusk, V. "The Syndrome of the Holy Degree: Critical Reflections on the Staff Development in Estonian Universities," *Higher Education Management*, 7, 3 (1995), pp. 385–397.

Tomusk, V. and Tomusk, A. "Teaching Psychology, Estonia: USSR Revisited," *Teaching of Psychology*, 20, 3 (1993), pp. 175–177.

Tse-Tung, Mao. *Quotations from Chairman Mao Tse Tung*, San Francisco: China Books and Periodicals, 1990.

US, *United States-Soviet Relations: 1988, Hearings Before the Subcommittee on the Europe and Middle East of the Committee on Foreign Affairs House of Representatives, One Hundredth Congress*, Vol. I. Washington: US Government Printing Office, 1988.

Vilakazi, H.W. "The Problem of Education in Africa." In: P. Higgs, N.C.G. Vakalisa, T.V. Mda, and N.T. Assié-Lumumba (Eds.), *African Voices in Education*, Lansdowne: Juta, 2000.

Wacquant, L.J.D. "Foreword." In: P. Bourdieu, *State Nobility*, Stanford: Polity Press, 1996.

Watson, A. *Legal Transplants: An Approach to Comparative Law*, second edition, Athens and London: The University of Georgia Press, 1993.

Wellmer, A. *Endgames: The Irreconcilable Nature of Modernity*, Cambridge: Massachusetts, and London, England: The MIT Press, 1998.

Winkler, H. "Funding of Higher Education in Germany." In: P. Hare (Ed.), *Structure and Financing of Higher Education in Russia, Ukraine and the EU*, London: Jessica Kingsley Publishers, 1997.

World Bank, *Constructing Knowledge Societies: New Challenges for Tertiary Education*, Washington, DC: The World Bank, 2002.

World Bank, *Constructing Knowledge Societies: New Challenges for Tertiary Education. A World Bank Strategy*, unpublished draft, Washington, DC: The World Bank, 2001.

World Bank, *Anticorruption in Transition: A Contribution to the Policy Debate*, Washington, DC: The World Bank, 2000.

World Bank, *Staff Appraisal Report: Russian Federation, Education Innovation Project. Report No. 16267-RU*, Washington, DC: The World Bank/Human Resources Operations Division, Country Department III, Europe and Central Asia Region, 1997.

World Bank, *Staff Appraisal Report: Romania, Reform of Higher Education and Research Project. Report No. 15525-RO*, Washington, DC: The World Bank/Human Resources Operations Division, Country Department I, Europe and Central Asia Region, 1996.

World Bank, *Staff Appraisal Report: Romania, Education Reform Project. Report No. 11931-RO*, Washington, DC: The World Bank/Human Resources Operations Division, Central and Southern Europe Departments, Europe and Central Asia Region, 1994.

Žižek, S. *Did Somebody Say Totalitarianism: Five Interventions in the (Mis)use of a Notion*, London, New York: Verso, 2001.

Index

academic (*see also* academician) 14, 50, 54–55, 62, 70, 73, 97, 98, 106, 107, 117–118, 123, 135, 145, 152, 154, 157, 161–165, 170, 189, 197–199
academic
  assessment 26
  autonomy 62, 112, 188–189
  capitalist 79
  colonialism 16, 156
  community 71, 165
  content 44, 62
  core 114–115, 118
  culture 89, 111
  entrepreneur 84
  excellence 33, 46, 77, 195
  freedom 63, 65, 72, 99
  goods and services 161
  heartland 117
  integrity 37, 76, 129, 162
  leadership 31–32
  leisure 144, 189
  mission 32, 192
  oligarchy 90–93
  profession 162–163, 177
  program 95, 101, 111, 114–115
  qualification 16, 77
  recognition 59, 115
  solidarity 84, 165
  staff 27, 71, 73
  standard 59, 71, 79–80, 192
  tourism 166
  traditions 141, 161
  training 125, 127–128
  tribes 107, 148, 153
  values 55
academician (*see also* academic) 79, 102

accreditation 20, 26–27, 29, 30, 43–46, 64–67, 79, 82, 86, 92, 95, 98–100, 108–109, 115–116, 119, 120–124, 128, 142
affirmative action 55, 137–139
Africa 137, 148, 151–152, 155–158, 160–163, 165, 170, 174
AINova (Academia Istropolitana Nova) 109–118
American University of Kyrgyzstan 79
Andropov, J. 19
Anthropic Principle 2
Apple, M.W. 86
Armenia 119
Arendt, H., Prize 89, 97, 110
Association of African Universities 151
attestation 65–67, 98, 122–123

bachelor degree 43, 59, 61, 69, 73, 82, 86, 103
Balcerowicz, L. 186–187
Bales, K. 153
Baltic States 19, 56, 77–78, 121, 190
Bandura, A. 9
Barrow, J.D. 13
Baudrillard, J. 89, 90, 94, 190, 192
Becher, T. 164
Belarus 19, 77, 90, 94–96, 98–103, 123, 142
Beria, L. 122
Berlin 54
Berlin, I. 135
Boia L. 32
Bologna Process 49–50, 62
Bolshevik 5, 15–18, 57, 134, 142, 199
Bourdieu, P. 84, 107, 131–133, 137, 141–143, 150–152, 197

brain drain 80, 151
Brennan, J. 124
Brezhnev, L.I. 19, 44, 54, 136, 196, 199
Broadfoot, P. 121
Bulgaria 83

Candidate of Science 59, 61, 103
capital 28, 57, 94, 131, 135–139, 142–143
  accumulation 3–5, 13, 137, 159–160
  cultural c. 131, 133, 136–137
  economic c. 81, 132
  exchange 137
  financial c. 58
  forms of c. 142, 170
  intellectual c. 118
  investment 69
  lackeys of c. 134–135
  social c. 107, 138, 143, 150, 173–174, 193
  symbolic c. 93, 133, 137, 142, 153
  human c. 134, 175, 181
  political c. 75, 112, 126
  types of c. 131, 133
capitalism 20, 54, 103, 127–128, 147, 152–153, 155, 158–159, 165, 197
Ceausesçu, N. 19, 21, 24, 26, 139–140, 142
center of excellence 114
Central Asia 15, 33, 78, 83, 79, 113, 122, 192, 196
CEU (Central European University) 110–111, 117
Chernenko, K.U. 19
China 158
choice 25, 29, 96, 99, 101, 147, 159, 161, 191, 200
  free c. 101
Chomsky, N. 135, 197, 200
Chuprunov, D. 138
CIS (Commonwealth of Independent States) 55, 59, 89
civil code 40–42, 47, 51, 63
Clark, B.R. 31, 75, 86, 90–91, 94, 102–103, 105–110, 116–117, 184
Clausewitz, C. von 8, 170
Colbeck, C. 107, 164

Cold War 118, 123
collectivization 15, 59
COMECON (Council for Mutual Economic Assistance) 54
Comenius University 93, 108, 111–112
commodification 126, 170, 173
Communism 3–4, 6–7, 15–17, 19–20, 54, 59, 76, 80, 84, 89, 134–135, 183
  Third Way C. 20
  Scientific C. 59, 82
  spell of C. 139
consumption 107, 149, 153, 155, 160–161, 166, 169, 172, 180–181, 197
  cultural c. 149
  levels of 148, 172
  logic of 197
  patterns of 148–149, 169
  private 4, 106
  standard of 169
  sustainable 180
CoE (Council of Europe) 59, 78–79, 86, 108, 121, 142, 190
control
  devolution of 63
  public c. 47
  bureaucratic c. 33, 54
  administrative c. 25, 40
  evolution of 63
  political c. 23
  central c. 6
  state c. 44, 63, 91, 107
Cuba 146
Cultural Revolution 59, 131, 158
Cyberspace 10
Czechoslovakia 110

Dahrendorf, R. 9, 90, 97, 111
Darvas, P. 79, 93, 141
Davies, J.L. 107
Davis, E. 10
Destatización 95
diploma mill 26, 28, 38, 76, 78, 87, 116, 141, 143, 198
donor 16, 78–79, 90, 96, 113–115, 118, 161, 194

agencies 161
country 78, 152
funding 79, 85, 114
single d. university 110
support 102
Drucker, P. 133, 191
Dutch model 120, 121

East Germany (*see also GDR*) 3, 6, 132, 183, 185
economic development 22–23, 53–54, 69, 145, 147–149, 158, 166
effectiveness 87, 96, 99, 120, 147, 196
Egorshin, A.P. 58
EHU (European Humanities University) 94–102
EIH (Estonian Institute of Humanities) 37
elite 14, 81, 93, 132–136, 139, 140–146, 155, 161, 168–169, 174, 179, 181, 190–194
 Communist e. 135, 137
 education 81, 143
 higher education 26, 165, 192–193
 national e. 162, 168
 new e. 57, 81, 139, 140–141
 political e. 155, 163, 189
 production 139, 141–142
 reproduction of 132, 134, 141
Elster, J. 184, 188, 189, 190
empiricism 8, 14
environment 85
Enlightenment 2, 17, 31, 151, 160, 200
enrolment 20, 24–26, 28, 55–56, 70, 76, 77, 146, 161, 165, 170, 187
entrepreneurial university 106, 108–110, 116–117, 124, 160, 178, 189
equity 96, 99, 101, 175,
Eritrea 179–180
Estonia 15, 20, 35–41, 43–51, 55, 84, 75, 77–78, 81, 95, 105, 120, 122, 128, 136
EU (European Union) 6, 22–23, 26, 27, 32–33, 37, 50, 75, 80, 82, 89, 114, 161, 163, 168, 189
European Commission 21–23, 91, 143, 163, 184, 193
evaluation 27, 122, 176

formative e. 122
procedures 177
summative e. 45, 122
Extropians 10, 11

faculty development 16, 192
FDI (Foreign Direct Investment) 168–169
financing 43, 65, 69, 96, 108, 111
formula funding 20, 43
France 132, 133, 138
Frederiks, M.M.H. 120
freedom 8, 24, 66, 72, 85, 96, 99–100, 102, 107, 115, 117, 124, 176, 200
 academic f. 63, 65, 72, 99
 of learning 200
 of research 72, 189
 of teaching 72
 of thought 199
 of training 72

Gaddy, C. 138
Gale, P. 157
GDR (German Democratic Republic, *see also East Germany*) 183–185, 193
Geiger, R.L. 110
Gellner, E. 106, 156, 157, 162, 172, 194
Gender Studies 100, 152
Gibbons, M. 164
Giddens, A. 138, 139
Globalization 108, 148, 152–154, 165, 168, 174, 176
Gorbachev, M.S. 19, 35, 37, 136, 139
Gorky, M. 57
GOSKOMVUZ (State Committee of Higher Education) 61, 125
Gosplan (Committee of State Planning) 60
graduate school 78
Graham, L.R. 55
Guaino, H. 172
GULAG (Glavnoe Upravlenie Lagerov) 6, 193

Hancock, G. 178–180
Hatar, V. 167
Havel, I. 11
Havel, V. 110, 140

Hayek, F.A. 17
Heclo, H. 14
HEEC (Higher Education Evaluation Council) 44
Hegel, G.W.F. 3
Higher Education Standard 80, 98
homo academicus 197
Honecker, E. 19
Humboldt, W. von 143, 188, 192–193
Humboldtian ideal 188, 192–193
Hungary 35, 77, 83, 109, 110, 121, 132, 137, 140–141, 175–177, 179, 194

Ickes, B. 138
Iliescu, I. 19
IMF(International Monetary Fund) 23, 167, 169
industrialization 15, 20, 24, 59
industry 3, 6, 16, 58, 60, 69, 71, 95, 127, 131, 133, 135–39, 146, 165, 186, 189, 194
Institute of Human Sciences 89
Internet 7, 9, 13–16, 153, 174
Ivory Tower 29, 38, 43, 154, 193, 198

Jackman, A. 9
Japan 24, 78, 150, 190–192, 195

Kant, I. 199
KGB (Committee of the State Security) 75, 184
Khrushchev, N.S. 136, 138–139
Kivinen, O. 81
kleptoklatura 140
knowledge
 academic k. 128, 157
 branches of 13
 commodification of 126, 170
 consumption 16
 economy 7, 32, 170
 indigenous k. 156, 158
 industry 134, 148, 165–166, 170, 181
 production 16, 17, 160, 164, 165, 166
 scientific k. 157
 society 161

symbols of 153
worker 198
Korea 118
 Peoples Republic of 159
Korka, M. 27, 30
Kuhn, T. 163
Kwiek, M. 197–198
Kyrgyzstan 79, 141

Lajos, T. 196
Lao-tzu 119
Latin America 31, 94, 102, 134, 137, 187
Leacock, S. 186
Leary, T. 94, 103
legal t. 35
legitimacy 57, 93, 94, 100, 105, 134, 141–143, 151, 161, 172, 184–185, 187, 191
Lem, S. 12
Lenin, V.I. 1–6, 15, 57, 126, 134–138, 196
Lennon, J. 174
Levy, D. 30–31, 78, 80, 94–99, 102, 134, 185, 187
Liberal Arts 37, 58
liberal education 45, 59, 115
licensing 58, 65–66, 77, 95, 98
Lilly, J. 12, 159
Lukashenko, A. 77, 100
Lumumba-Kasongo, T. 155
Lysenko, T. 195

Makgoba, M.W. 148
Mao Tse-tung 131
market
 academic m. 153
 capitalism 153, 159, 165
 coordination 75
 economy 22, 28, 165, 171, 189
 educational m. 16
 global m. 8, 21, 126, 168, 171
 labor m. 32, 38, 59–61, 78, 82–83, 145, 150, 192, 195
 mechanism 22, 29, 90
 reforms 75, 77, 82–83, 95, 186
Marx, K. 2–4, 7, 10, 13, 17, 33, 57, 135, 196

Marxism 3, 135
Marxism-Leninism 37, 135, 197
Matos, N. 151, 158
Matthews, M. 139
MBA (Master of Business Administration) 16, 37, 134, 163, 180, 200
McLuhan, M. 94, 153, 168
Merton, R.K. 4
Middle Ages 2
Mitchurin, I.V. 106, 195
Mokhov, V.P. 136, 137, 139
Moscow School of Social and Economic Sciences 85
Moura Castro, C. de 31, 187
Moynihan, D.P. 172
Murakami, H. 10

Naan, G. 136
nanotechnology 9
NATO (North Atlantic Treaty Organization) 23, 132
NCAEA (National Council for Academic Evaluation and Accreditation) 26–27, 30, 122
Neave, G. 75, 83, 106–107, 200
NEP (New Economic Policy) 57
Neumann, I.B. 132
Newton-Smith, W.H. 13, 197
Nkomo, M. 156–157, 160
nobility 107, 125, 131, 134, 136–137, 141–143, 149
nomenklatura 76, 81, 132, 140, 183, 190
numerus clausus 83, 126

OECD (Organization of Economic Cooperation and Development) 25, 49, 55–56, 59, 86, 107, 114, 126, 132, 145–146, 153, 165, 170–172, 178, 189, 194
Offe, C. 153, 159, 188–190, 192
Open University 85, 175

Perestroika 35, 37, 136
Petrova, I.E. 70
Ph.D. 60–61, 115, 165
Pirsig, R.M. 124

Polányi, M. 157
Popper, K.R. 12, 152
Popperismus 185
Potanin, V. 140
private higher education 25–27, 30, 33, 37–40, 42, 44–46, 56, 58, 63, 76–78, 82, 92, 94–96, 108, 116–117, 119–120, 191
private university 25, 27–30, 33, 46, 56, 64, 71, 75–76, 80, 82–84, 86–87, 95–97, 99, 100, 102, 108–109, 114, 121, 141, 185
privatization 47, 81, 95, 139–140, 143, 193
proletariat 5, 126, 127, 137, 140, 143, 197
public good 14, 106, 113
Putnam, H. 124

quality assurance 20, 26, 30, 43–45, 65, 78, 98, 119–124, 129, 142, 190

Ramonet, I. 168–169
Ratcliff, J.L. 119–120, 123
Readings, B. 108, 154–155, 161
relevance of higher education
  economic r. 72
  intellectual r. 147
  local r. 155, 161, 171
  political r. 161, 163
  social r. 165
  substantive r. 161, 163
reproduction 107, 131–132, 134, 141, 147
Rév, I. 183
Rinne, R. 81
Rodham Clinton, H. 79
Rogalina, N. 4
Romania 15, 19–33, 46, 50, 55, 65, 77, 86, 108, 119–122, 139, 141–142, 178–179, 195
Roper S.D. 139
Rorty, R. 196
Russia 4–6, 15, 20, 22, 53–54, 56, 58–63, 65–66, 69–72, 76–77, 98, 108–109, 121, 134, 140, 195–196
Ryzhkov, N. 136

Sadlak, J.   19, 23, 25, 26
Salmi, J.   174
Sapatoru, D.   23, 25
Savchuk, V.   141
Schutz, A.   147
Seepe, S.   156, 158
self-fulfilling prophecy   4
Senegal   175–177
Shevardnadze, E.   136
shock therapy   20, 75, 187
Siddhartha Gautama (Buddha)   2
Skillbeck, M.   107
Skinner, B.F.   181
Sklair, L.   168–170
slave   5–6, 54, 153, 193
Slovak Republic (*see also Slovakia*)   110–111, 113
Slovakia (*see also Slovak Republic*)   108, 110–111, 113–115, 118
Slovenia   193
social
  justice   106, 138, 148
  mobility   26, 106–107, 131, 133, 149–150, 187, 188, 193
  science   16–18, 59, 73, 79, 101, 110, 115, 127, 146–147, 149, 152, 163, 166, 190, 194
socialism   35, 158
  state-s.   4, 6, 47, 53, 135, 159, 185
Socrates   83
Sokolov A. K.   58
Soros, G.   90, 110–111
Soviet
  domination   56
  higher education   36, 43, 53–55, 58, 126, 146
  intellectuals   54
  leadership   21, 73
  legislation   38
  period   36, 49
  regime   37
  society   5
  system   29, 42, 53–54, 57, 60, 102, 135
  universities   57, 73
Soviet Union   5–6, 15, 19, 20, 33, 35–37, 40, 44, 47, 53–60, 62–64, 67, 73, 75, 77–81, 83, 94, 96, 102, 116, 119, 121–122, 125, 131–132, 134–135, 138, 140–142, 146, 149, 158–159, 183–184, 188, 190, 192–194, 199
Sovnarkom (Council of Peoples' Commissars)   18
Stalin, J.V.   3, 20, 53–54, 56–57, 84, 136–137, 196, 199
State Educational Standard   58, 61, 64, 67–68
state socialism   4, 6, 47, 53, 135, 159, 185
Szántó, A.   14
Sztompka, P.   154

Tannehill, M and L.   150
TCC (Transnational Capitalist Class)   168–170, 176, 180
Theory of Everything   13, 154, 163
Tiazhel'nikova, V.S.   58
Tiger Leap   15
Tipler F. J.   13
TNC (Transnational Corporation)   168–169, 177, 180
Todd, E.   4, 35, 135
transitology   190
transplant   20, 35, 75, 79, 94
Trotsky, L.   2, 196

United Kingdom   5, 20, 146
University of Tartu   38–39, 41–43, 48
University of Vilnius   83
utopia   2–3, 6–8, 10–13, 15, 17, 35, 54, 75, 106, 145, 172, 181, 192
  digital u.   10–11, 106

Van Vught, F.   107
Vietnam   118, 171
VHEI (Vocational Higher Education Institution)   41–42, 44
Vilakazi, H.W.   152, 158
vocational higher education   39–42, 48, 58–59, 64, 77, 86, 124, 132
Voznessenski, A.   53
VSNU (Association of Universities in the Netherlands)   120

Watson, A.   40, 91
Western Europe   2, 66, 108, 154
westernization   45, 119, 155, 161, 189–190, 192, 198

World Bank 7, 33, 78, 86, 145, 155, 160, 167, 169–171, 173–181, 190, 193–194

World War II 8, 20, 53, 56, 150, 172, 190–192

WTO (World Trade Organization) 157, 161, 167

Yugoslavia 21, 162

Žižek, S. 197

GPSR Compliance

The European Union's (EU) General Product Safety Regulation (GPSR) is a set of rules that requires consumer products to be safe and our obligations to ensure this.

If you have any concerns about our products, you can contact us on

ProductSafety@springernature.com

In case Publisher is established outside the EU, the EU authorized representative is:

Springer Nature Customer Service Center GmbH
Europaplatz 3
69115 Heidelberg, Germany

www.ingramcontent.com/pod-product-compliance
Lightning Source LLC
LaVergne TN
LVHW041626060526
838200LV00040B/1452